WOOD

IN

CONSTRUCTION

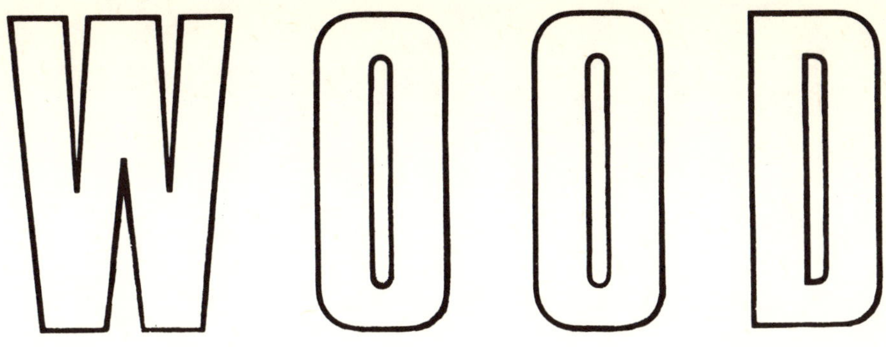

WOOD IN CONSTRUCTION

Barry A Richardson

Director of Penarth Research Centre

THE CONSTRUCTION PRESS LTD

To Janet, my wife

who tolerates months of household disruption whenever I decide to write a book

ISBN 0 904406 14 8

Published in 1976 by:

The Construction Press Ltd.,
Lunesdale House,
Hornby,
Lancaster, LA2 8NB.

Printed in Great Britain by
The Blackburn Times Press

Contents

Preface

Have you ever wondered why a particular wood may be associated with a specific function? For example, greenheart is favoured for marine piles and coniferous softwoods are used for joists, rafters and other load-bearing components in buildings. In boats teak or mahogany may be used for planking, yet oak for frames and spruce for masts. Scots pine is better than maritime pine for pit-props, whilst the spruce favoured for boat masts is rarely used for transmission poles which must be pressure impregnated with preservatives. Obviously the traditional uses for wood have developed as a result of experience but the suitability for a particular use is directly related to the properties of the wood. As traditional species and qualities become less readily available an understanding of these essential properties becomes increasingly important in the selection of alternative species.

Anyone who has attempted a study of wood science and technology will have found that the few available books are written largely for the benefit of the specialist, usually reflecting the enthusiasm of the author for particular aspects of the subject. Most of the authors are distinguished botanists and their books usually contain extensive sections on microscopic features and identification which are of very limited interest to the practical user or specifier of wood in construction. In addition these highly technical books frequently refer extensively to published papers in order to justify comments and conclusions but, whilst this is correct scientific practice, it is again of little value to the user, specifier and even the wood technology student; this information is only of interest to the expert or research reader. Simplified statements of fact are available, usually in the form of booklets on specific aspects of wood science and technology published by government departments and technical associations throughout the world which are attempting to encourage efficient wood utilization. Whilst these sources are very valuable there is clearly a need to collect the basic information so that it is readily available, accompanied by the explanations that are a necessary part of such a handbook if it is to be of value to intelligent students, specifiers and users of wood for structural purposes.

"Wood in Construction" is an attempt to provide the basic information and explanations that may be required to ensure the intelligent utilization of wood for structural purposes. Whilst the use of wood for load-bearing components in buildings is perhaps the most important subject covered by this book there are many minor uses described which perhaps involve wood technology to a greater extent. The book is written to cover the uses of wood in the temperate and neighbouring zones of the world. Tropical woods are included if they are utilised in these zones but not if they are 'secondary species' utilised only locally in the tropics. Care has been taken to avoid conflicting terminology; for example, the word 'timber' with different interpretations in English and American is not used and the English term 'joinery' is always followed by the American 'millwork'.

Whilst "Wood in Construction" will provide the architect, engineer and general wood technology student with a valuable reference volume there is clearly a need for more detailed and more comprehensive information on particular aspects of wood science. This will be provided in the form of further volumes in this series; "The Strength Properties of Timber" has already been published but "Wood Preservation", "Wood Engineering" and "Wood Identification" will also be available shortly. In addition it is planned to cover other construction materials in a similar way; "Stone in Construction" will be the next general volume.

Finally, I wish to acknowledge the considerable help that I have received from my colleagues and friends. In particular I must thank the Princes Risborough Laboratory, UAC Timber, Finnish Plywood Development Association, Timber Research and Development Association, Council of Forest Industries of British Columbia, Anticimexbolagen, Wykamol Limited, Richardson and Starling Limited, Cook Bolinders Limited and Orban Bois S.A. for supplying many of the illustrations, which always make a technical book of this type so much more interesting.

Barry A. Richardson

Penarth House,
Otterbourne Hill,
Winchester

August, 1975

1. Wood as a Material

Introduction

The twentieth century has already seen incredible technological advances, quite unparalleled in the history of the human race, some of which have resulted in profound changes in the construction industry. New materials have been introduced, new manufacturing methods have enabled traditional materials to be used more efficiently or more cheaply, and the demand for other traditional materials has declined. Natural stone masonry has fallen from fashion because it is said that it is expensive, inefficient or perhaps simply non-modern. The use of wood for structural purposes has been similarly criticised and it was forecast only a few years ago that it would eventually be eliminated from all construction uses to be replaced by the more modern, more uniform, manufactured materials such as concrete, metals and plastics. In fact, these dismal forecasts of only a short time ago can now be seen to be fundamentally incorrect. The decline has been halted and there is a resurgence in the use of wood for structural purposes so that it is progressively replacing its competitors in such widely divergent fields as piles, poles, furniture, window frames and concrete shutterings. Clearly there should be some dramatic explanation, perhaps some fundamental error, that explains the incorrect forecast of the recent past but no single factor can account for this situation which arises through the many factors that account for the various advantages of wood.

Merits of wood

Anyone who has used wood, whether in his own home workshop or in some enormous industrial production process, will be aware of the simplicity with which it can be worked by hand or by machine. It is, in fact, incredibly simple to fabricate structures from wood and, even in the most sophisticated production processes, the tooling costs are relatively low compared with competitive materials. Wood is ideal if it is necessary to erect an individual structure for a particular purpose but it is equally suitable for small batch or mass production, particularly now that finger jointing processes have largely eliminated the wastage that previously arose when cutting to length. When these working properties are combined with the other advantages of wood, such as its high strength to weight ratio, its excellent thermal insulation and fire resistance, and the unique aesthetic properties of finished wood, it sometimes becomes difficult to understand why alternative materials have ever been considered! However, there is one feature of wood which is unique amongst all structural materials; it is a crop which can be farmed whereas its competitors such as stone, brick, metal and plastic are all derived from exhaustible resources. This feature is alone sufficient to ensure that wood will continue in use as a structural material virtually for ever and it is also likely to ensure that wood remains the cheapest of all structural materials.

1.1 Wood technology

Wood is, of course, the natural supporting skeleton of the larger plants and it is important to understand its origin in order to fully understand its properties. In simple terms a plant consists of roots, stems and leaves, and a tree is only different from other plants in the scale of its development. A massive crown of leaves is developed which is only efficient if it can avoid being overshadowed by neighbouring plants. Thus a massive stem or trunk is developed which will support this crown of leaves at a height sufficient to ensure the continuing growth and survival of the plant. In turn a massive root system develops in order to supply the entire plant with water and minerals, and to anchor the plant to the soil in which it grows. Wood is simply the skeleton that is developed within the trunk in order to support this massive structure. In fact the trunk does not perform a passive supporting function alone but also acts as a storage area, whilst the outer zones provide the conducting channels through which materials are conveyed between the root and the crown, as well as performing the growth function which ensures the enlargement of the trunk to enable it to support the progressively growing plant.

As a plant a tree consists of a crown of leaves, generally supported on a single main stem known as the trunk or bole which connects the crown to the roots in the ground, a growth arrangement that is known as the dendroid habit. The sole purpose of this very elaborate structure is simply to survive and supply the cells within the plant. This is achieved firstly by the roots which absorb water, containing dissolved mineral salts, which is then conveyed by the trunk, branches and twigs to the crown and the individual leaves. The function of these leaves is to absorb carbon dioxide which is combined with the water from the roots to form simple sugars by the process

Figure 1.1 **The living tree and photosynthesis.**

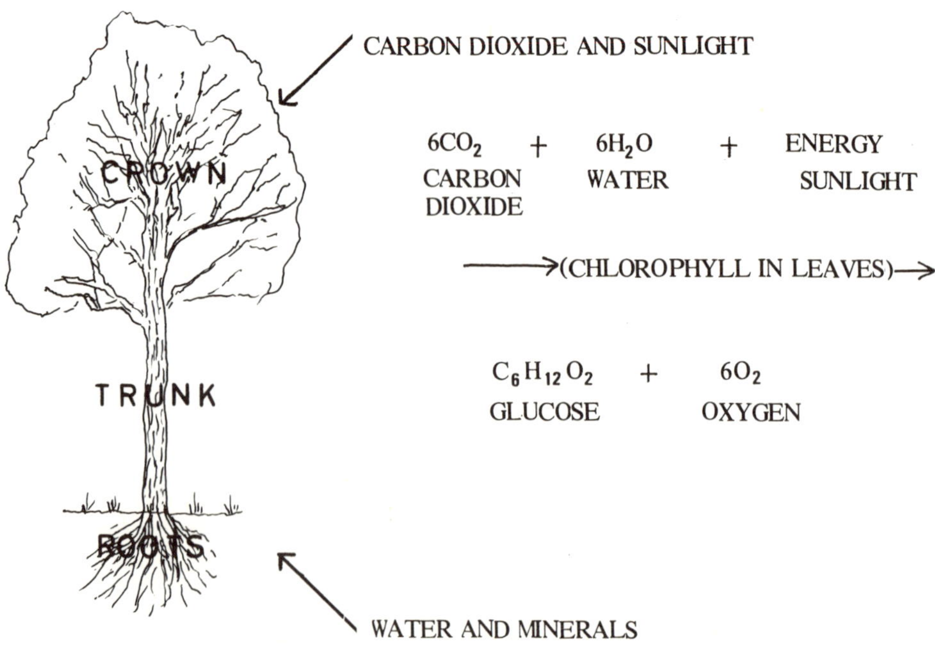

known as photosynthesis; the cholorophyll in green plants is the essential catalyst which enables this process to proceed whenever there is adequate ultra-violet radiation received from the sun. The sugars are then conveyed throughout the plant, to the leaves, twigs, branches, trunk and roots. The primary function of the sugar is to provide an energy source, or food, for the individual living cells within all these components of the tree but a secondary function is to provide the basic units from which the skeletal structure of the tree is formed. In fact, whenever there is sufficient sunlight simple sugars will be produced by the leaves and will be found distributed throughout the living tissues of the tree. Some of the sugar will be formed into starch deposited within the living tissue as a reserve energy source which can be utilised by the cells whenever sugar is unavailable from the leaves, such as at night or during winter months when deciduous trees shed their leaves. Finally the sugar units are linked into cellulose chains which are then assembled into the main skeleton of the woody parts of the tree.

Wood formation

There are certain aspects of this basic structure that should be described in greater detail. A tree, of course, is a perennial plant but, whereas the herbaceous perennials lose their aerial parts annually and survive only in equable soil, a tree possesses permanent aerial parts and permanent embryonic apices to the stems and roots which enable growth to recommence from these points, even though the tree may shed leaves or even some branches during the seasons of the year which are unsuitable for growth. It is these properties of the tree that enable it to withstand extreme winter cold and summer drought, or temporary destruction by plant and animal pests. Even an individual leaf has been developed so that it conserves water during summer heat; the stomata on the underside of the leaf through which water loss principally occurs are each surrounded by a guard cell which reduces the aperture in size during bright sunlight.

Tree growth

A tree is, of course, forever increasing in size and this is the function of embryonic tissue distributed around the whole plant. The increase in the crown is the responsibility of the primary meristem, essentially the bud tissue at the end of each twig. The centre of the twig is the pith which terminates in the promeristem or apical meristem which is a little mass of tightly packed cubicle cells with abundant protoplasm and which are capable of very active division to produce similar cells from which the new material is formed and which thus enables the twig to increase in length. The detailed structure of the bud and this apical meristem has little significance to the use of wood in construction but its importance lies in the fact that this meristem tissue extends over the entire surface of the tree, just beneath the bark of the twigs, branches and trunk and extending similarly over the entire root system. The purpose of this secondary or lateral meristem is to enable all these structural components of the tree to increase laterally or in girth so that they are capable of supporting the enlarged crown of the tree. Each twig as it lengthens consists initially only of pith formed by the primary meristem but it is covered externally by the secondary meristem to permit subsequent increase in girth, as well as a protective epidermis to control water loss and to prevent disease damage. As the meristem tissue generates new cells which increase the girth this epidermis splits, exposing inner tissue, but the meristem tissue then generates a new protective layer which becomes the rough outer bark of the branches and trunk.

Usually a trunk in its simplest form would consist of a single twig, progressively increasing in length or height as the twig is extended by the

Figure 1.2 Diagrammatic trunk showing annual rings.

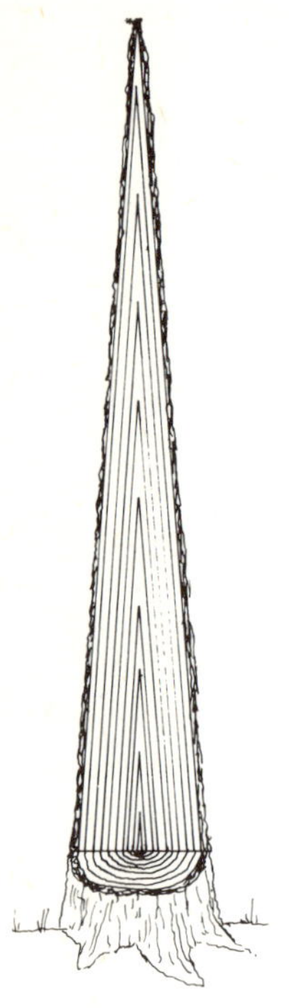

primary meristem at its apex. In order to support this increase in height the whole of the trunk is covered in secondary meristem in order to achieve the necessary increase in girth so that the trunk possesses a steep conical shape which is essentially the portion of the tree which is the wood of commerce. It consists of a central pith enclosed, in effect, by a series of cones, each cone representing the annual growth increment or an annual ring. The wood tissue round the pith is the heartwood and consists of dead cells. The heartwood is surrounded by the living cells of the sapwood or xylem which is, in turn, covered by a thin layer of phloem cells and finally the protective bark. The interface between the xylem and phloem cells is known as the cambium which is the term given by wood technologists to the actively dividing cells which have been previously described as the secondary or lateral meristem. The actively dividing cambium cells are known as fusiform initials, the cells splitting off on the inner side of the cambium becoming xylem or wood tissue and those on the outer side of the cambium forming phloem or bark tissue. In coniferous trees the xylem cells are termed tracheids whilst in dicotyledons or broad leaved trees they are termed fibres. The cambium also possesses additional active cells known as ray initials which generate horizontal radial rays of parenchyma storage tissue within the xylem.

Spring and summer wood

This description is not very complex, yet it explains the growth structure of wood within the trunk of a tree. Xylem can only be formed when the entire tree structure is active. The xylem deposits thus tend to be seasonal, particularly in temperate areas, so that a trunk will increase each year by the addition of an outer cone of tissue representing the growth for a season. This growth varies from a wide band of large but thin walled cells, the early or spring growth, to much smaller thick walled cells, the summer or autumn growth. This late growth terminates sharply as cell formation discontinues with the onset of winter and this sharp terminal line is then followed by the large cells of the early growth for the following season.

Wood cell

These details explain the basic structure of wood but they have additional significance. The xylem or sapwood is the tissue that conducts water and dissolved salts from the roots to the crown of the tree whilst the phloem is the tissue that conducts sugars from the crown to the various growing cells throughout the structure of the tree. When a xylem cell, a tracheid or fibre, is first formed it consists of a thin wall of sugars which have polymerised into cellulose. The sugars from the phloem continue to be supplied to xylem cells, the rays perhaps providing routes for this transfer, so that successive secondary layers of cellulose are formed within the original primary wall. Once the cell structure is complete, a relatively rapid process occurring within the original growth season, the much slower process of lignification commences. This consists of the progressive deposition of lignin, initially within the middle lamella, an amorphous undifferentiated region between the cells, but then within the cellulose cell walls. This lignification serves to stiffen and strengthen the cells, occurring progressively whilst the cells remain alive. The sapwood or living xylem cells consist of a reasonably constant number of rings or depth of xylem, presumably controlled by the availability of food and oxygen to maintain the living cell processes, so that a further annual growth at the cambium is accompanied at the inner surface of the sapwood by the death of cells and their conversion into heartwood. The only significant difference between sapwood and heartwood is the large amount of material that is deposited in the latter, apparently waste material arising from the living processes of the tree. These deposits in the heartwood cells are often significantly toxic so that heartwood is generally more resistant to insect and fungal attack than sapwood. These deposits also tend to make heartwood more stable so that it is much more resistant to swelling and shrinkage with changes in moisture content.

Figure 1.3 **Wood zones in a trunk.**

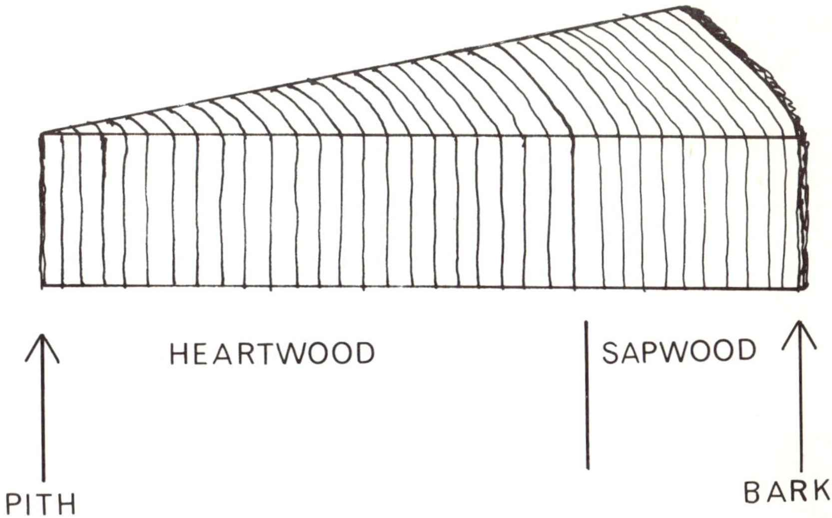

Branches and knots

Obviously the trunk of a tree is not a simple twig but also possesses branches. Typically the primary or apical meristem at the end of the twig divides to form several additional buds which in turn give rise to separate twigs. In many coniferous trees this side budding occurs annually so that examination of the trunk of the tree discloses a ring or whorl of side branches. The distance between individual whorls signifies an annual height increment. When the trunk is sawn to provide the wood of commerce these side branches are related to knots, the roots by which the side branches penetrate through the main tissue of the trunk. If the branch is removed at some stage, either accidentally or deliberately, the knot becomes blind and completely covered by subsequent growth of the trunk.

Botanical classification

The basic properties of a piece of wood depend upon the species of the tree from which it is derived. Whilst this book is concerned with wood in construction some knowledge of botanical classification is essential, and Appendix 1 gives a classification of the principal north temperate timber trees, showing how the plant kingdom is divided into four principal divisions, one being the Spermatophytae or flowering plants. This is in turn divided into two sub-divisions, the Gymnospermae which comprises the order Coniferae, the cone-bearing plants or commercial softwoods. The second sub-division is the Angiospermae which includes the order Monocotyledonae, the palms and grasses, and the order Dicotyledonae, the broad-leaved plants or commercial hardwoods. The palms are rarely used as sources of wood for structural purposes so that the conifers and dicotyledons form the two main groups of timber trees.

Appendix 1 shows how the Coniferae are divided into three families, the principal being the Pinaceae and the best known genus in this family being *Pinus*. Within this genus four different species are shown, *Pinus strobus, Pinus sylvestris, Pinus nigra* and *Pinus pinaster* but there are, of course, other species that have been omitted from this short list of examples. The purpose of this classification is, of course, to establish the various relationships between the plants and thus between the woods of commerce. Obviously the pines, genus *Pinus,* are reasonably similar to each other in basic structure but they are distinctly different from, for example, the larches, genus *Larix,* and even further removed in structure from the cypresses, genus *Cupressus.* In many respects a close botanical relationship will mean reasonably similar physical properties and woods that are suitable for similar uses. The other use of the botanical classification is identification; a sample might be rapidly identified as a conifer whilst close examination with a hand lens and microscope will ultimately enable its species to be identified. The dicotyledons, or commercial hardwoods, are similarly classified, although they involve a rather larger number of families than for the softwoods.

Softwoods and hardwoods

There are fundamental differences in the nature of the wood of the Coniferae or softwoods and the Dicotyledonae or hardwoods. In softwoods the principal longitudinal cells are known as tracheids which serve both conducting and mechanical support functions. In transverse section a piece of wood appears as a honeycomb with the annual rings arising through the change in density of appearance between the early wood with its large thin walled cells and the late wood with its small thick walled cells. The transverse section may also show occasional vertical resin canals whilst the radial and tangential sections may show horizontal features such as rays. In contrast the hardwoods possess fibres, similar to softwood tracheids, to provide structural support but the conducting cells are termed vessels or pores. These vessels are distributed singly or in clusters throughout the wood, or in radially or tangentially orientated groups. For example, a tangential distribution results in the ring porous woods, a term that results from the observation that a distinct ring can be seen with the naked eye on a transverse section of species of this type such as oak, ash or elm. Whilst the transverse and radial sections of both hardwoods and softwoods will generally show annual rings which may appear to be superficially similar, they actually result in rather different variations in the structure; these differences are described in detail in Appendix 2.

Transverse, tangential and radial section

The microscopic features that are apparent in transverse, tangential and radial sections naturally contribute to the general appearance and properties of a piece of wood. The structural elements, the fibres and tracheids, are longitudinally orientated so that a cross-cut will disclose a porous structure with the annual rings clearly apparent. A radial cut will also disclose the annual rings but as parallel longitudinal lines. Whilst a strictly tangential cut will obviously follow the curve of an annual ring a commercial flat sawn board will cut across the rings, so that they appear towards either edge as relatively straight lines similar to a radial cut and their irregularity resulting in loops of rings towards the centre of the board. It is not intended to discuss the methods of converting logs to boards at this stage but the direction of cut is important when considering the various terms that are used to describe the appearance of wood. A quarter sawn board follows one of the radii of the tree. Flat sawn boards are derived from a simple slicing of the trunk and, with the exception of the centre board which is a radial cut, all other boards are tangential at their centre but with the rings emerging from the face at a progressively steepening angle towards the edges.

Quarter and flat sawn boards

Grain, texture and figure

Wood is generally described in terms of its grain, texture and figure. These features are often confused but grain should refer to the direction of the fibres or tracheids, texture to the relative size and amount of variation in the structural elements, and figure to the appearance generated by natural variations. Before discussing these features in detail it is necessary to comment upon the influence of the environment upon a tree. In good conditions with ample light, water, minerals and an equable climate, the tree will generate a rapidly expanding crown of leaves and there will be a substantial sugar production which will permit the formation of wide annual rings. In less suitable conditions, such as temperate trees growing close to the Poles or at high altitude, the rate of production of sugar will be less and the rings narrower as a result. This is, of course, a generalisation as ring width is also substantially influenced by the planting density. It is quite common to observe in a cross-section of a trunk that the annual rings tend to become narrower towards the bark but there is often a sharp change to much wider rings as a result of thinning allowing more light to the crown. Scots pine grown in England has a ring density of perhaps 10–20 per inch (4–8 per cm). The same species grown in the Baltic may have twice this density and the ring width decreases progressively as trees grow further north. The wide rings of rapidly grown wood generally mean low density, perhaps a lack of stiffness and high movement, i.e. excessive shrinkage on drying which may lead to warping and other distortion defects. Rapidly grown wide-ringed wood is also more difficult to work as it tends to splinter during cutting and planing. The quality in all these respects is greatly improved as the ring density increases, for example from English Scots pine to Baltic Scots pine (redwood). However, a move further north results in much closer rings but, whilst the wood is "sweeter" to work, it is not so strong. Thus Archangel larch is particularly suitable for joinery manufacture where workability is more important than absolute strength. Whilst thinning allows more light to the head and thus wider rings, the same phenomenon occurs in trees at the edge of a forest; the exposed side receives more light, producing wider rings and thus an eccentric growth pattern.

Ring density

Each annual growth increment consists of a ring or, more correctly, a cone of new wood composed principally of structural tracheids in softwoods or fibres in hardwoods. If this growth ring is observed from outside the trunk, so that a tangential section is studied, it will be found that these tracheids or fibres do not invariably run in a vertical direction, or longitudinally within a piece of wood, and this gives rise to the characteristic known as grain. In a straight grained piece of wood the structural elements were vertical in the tree and, after intelligent conversion, should be longitudinal within the final piece of wood, giving good strength, easy working but a rather unattractive appearance. Diagonal or cross grain is the description given to a piece of wood in which the annual rings or the grain in the tangential section are not parallel to the cut edge, introducing difficulties in sawing and particularly in planing. Any slope of the annual rings or the grain introduces a weakening in a piece of wood so that a slope of 1 in 10 reduces the bending strength by 20% and at 1 in 5 the loss is increased to 50%; for beams the slope should never exceed 1 in 15. This is one of the reasons why knots should generally be excluded from load bearing members as they introduce distortion of the grain and thus a weakness.

Diagonal or cross grain

Spiral or torse grain

Spiral or torse grain is the description given to the situation when the tracheids or fibres within an annual ring follow a steep spiral around the tree. This phenomenon is particularly common in Scots pine and occurs occasionally in spruce and fir. It can sometimes be identified in the standing tree by the presence of spiral cracks and fissures in the bark as the structure of the bark follows the structure of the xylem beneath. In sawn wood spiral grain sometimes becomes apparent in the form of diagonal seasoning cracks but it is difficult to detect in the absence of these obvious defects; a drop of ink can be used to detect the grain as it will spread in the direction of the tracheids or fibres.

Interlocking grain

In interlocking grain the slope of the spiral varies in direction. When facing the trunk of the tree several successive rings might spiral to the left but a few rings later the direction might be to the right. Thus a radial cut exposes the different reflectivity arising through the different angle of the fibres and tracheids to the cut surface, producing light or dark bands and "stripe" or "ribbon" figure. This is common in many tropical woods such as the African mahoganies, sapele and Grand Bassam. Interlocking grain invariably causes planing difficulties as one stripe will lie smooth whilst the adjacent stripe will lift. Woods with interlocked grain are also very difficult to split in the radial plane, although this can be both an advantage and a disadvantage, depending upon the situation. Heavily interlocked woods, perhaps 20–30° from vertical and with changes every $\frac{1}{4}$–$\frac{1}{2}$″ (0.5–1.0 cm) along a radius, as in keledang, sepul and terentang, can cause considerable sawing difficulties as the fibres tend to wrap around the teeth and jam the blade.

Plate 1.1 Stripe figure in sapele (radial cut).

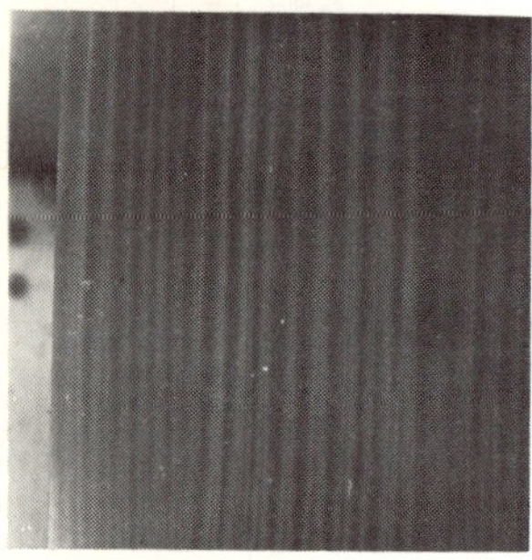

Sinuous or wavy grain

The grain may be irregular or wild, indicating that the fibres change direction in patches at random, making the wood extremely difficult to work and doubtful in strength. Sinuous or wavy grain is caused by undulations in the fibres over a length of $\frac{1}{4}$– 1″ (0.5–2.5 cm). Wavy grain often occurs in oak, ash, horse chestnut and sycamore, the latter being highly prized for the backs of violins and 'cellos where it is known as fiddleback mottle; when varnished a slight rocking of the surface will give beautiful changes of light and shade. Wavy grain considerably weakens the wood but can give a number of interesting effects. Ramshorn is a special form of wavy grain with short waves whilst feather curl is an ostrich feather pattern, sometimes seen in crotch mahogany. Roe figure is a combination of wavy and interlocking grain to give a beautiful broken ripple. Yet another form of irregular grain is known as blister grain and is caused by irregularities in the cambium. When Douglas fir is flat sawn or prepared as a peeled veneer the changes in the tracheid direction give a blister appearance to the surface of the wood. The summer wood is denser and darker but stains are absorbed preferentially by the softer spring wood so that a variety of contrast effects can be produced. Where the tracheids are dimpled inwards in some hardwoods a smaller type of blister grain can be produced, as in burr maple which is so highly prized for panelling. Sometimes these effects are caused when the cambium is irritated by some outside influence, such as a fungal attack which causes large burrs at the base of a tree or where pollarding causes a large burr to be formed where the twigs are cut; if twig knots are apparent the effect is known as bird's eye figure. There are other features that are often termed grain but as they are not caused by tracheid or fibre direction they are more correctly described as figure and will be considered later.

Fiddleback mottle

Ramshorn, feather curl and roe figure

Blister grain

Burr and bird's eye figure

Texture

In the structural sense grain is certainly the most significant feature except for obvious massive defects such as knots and inclusions. Texture is, however, very important as it is concerned with the uniformity of the wood and arises through the relative sizes and the amount of variation in the structural elements. Descriptions of texture often lead to confusion because of the fundamental differences in the structure of softwoods and hardwoods. In hardwoods a coarse textured wood indicates large vessels and, as the vessels are largely responsible for the contrast between spring and summer wood, it is very difficult to distinguish the annual rings in a very fine textured wood such as box. Fine textured hardwoods are very easy to saw, plane and sand to a good finish. In softwoods the texture arises through the contrast between the spring and summer wood which generates the annual rings. If the rings are uniform in width the wood is even textured, or incorrectly, even grained. If the rings are close together the wood is considered to be fine textured or close grained, although the best term is probably dense ringed. If a softwood is considered to be coarse this generally indicates sharply contrasting annual rings, as in Pitch pine where the summer wood is particularly dense and prominent, yet the even texture of another conifer, Virginian juniper, makes it easy to work and particularly suitable for pencil manufacture; it is often incorrectly called pencil cedar.

Figure

Figure is the term that is used to describe the appearance of wood in the decorative sense, whatever its cause. Grain is only appropriate where a feature results from the orientation of the principal longitudinal structural elements, the tracheids or fibres, but it is sometimes incorrectly used to describe various features arising from other causes. For example, silver grain should be termed silver figure as it arises through the distribution of the medullary rays in a quarter sawn piece of wood. It commonly occurs in oak, plane and beech where the rays are deep, up to 1″ (2.5 cm) in oak, giving bright irregular splashes against the darker background of the structural elements. Oak cleaves down the rays in the radial plane, showing silver figure to its best advantage, and cleft "feather edged" panelling boards with beautiful silver figure can still be found in old country houses. Silver figure caused by the rays gives the characteristic appearance of Australian silky oak. In some species the parenchyma tissue of the rays is apparent on flat sawn boards, for example in *Millettia* spp. where it gives characteristic flame or watered silk figure.

Flame or watered silk figure

Crotch or curl figure

In crotch or curl figure the leading shoot of the tree was damaged and two new shoots subsequently developed to produce a crotch. Slices across the crotch may produce the very valuable curl figure veneer, a particular feature of rosewood.

Brittleness

Wood that suffers brittle fracture is sometimes described as being "short in the grain" but this is an inappropriate term as it is really due to the length of the structural elements. In some cases the brittleness is an inherent property of a particular species or it may exceptionally arise when a specimen has a rather lower density than normal. Brittleness is sometimes associated with compression wood, the heavier growth that occurs under branches. Tropical woods are usually more brittle than temperate woods, largely because of their higher lignin content, but they are also stiffer and more suitable for many purposes. Brittleness is also commonly caused by fungal decay or by incorrect seasoning involving too rapid drying at high temperature and low humidity.

Colour

Colour is also an important feature with considerable influence over choice for a particular purpose. Ebony and sycamore are selected for decorative purposes because of their dark and light colours respectively, whilst mahogany and walnut are chosen because they combine attractive colour with interesting figure. Colour can be generally attributed to the infiltrates which are deposited within the heartwood as part of the death process which distinguishes the heartwood from the sapwood. Thus the heartwood is almost invariably darker in colour than sapwood, except in a few mild and light coloured woods such as beech, sycamore and spruce where the sapwood and heartwood are almost indistinguishable. Generally the colour of an exposed wood surface will darken with age, largely as a result of the oxidation of some of the infiltrates, and this darkening will occur more rapidly if exposed to light. Exposure to weather invariably results ultimately in the loss of the original colour and the development of a grey colour, largely because many of the infiltrates have been leached from the surface by rainfall; in a building clad with western red cedar the colour is often retained in the protected areas under the eaves.

It is sometimes suggested that the infiltrates deposited within the heartwood represent toxic materials which the tree wishes to eliminate from the living cells. Certainly darker coloured heartwoods appear to be more resistant to the attack of wood destroying insects and fungi, although the resistance to fungi may not be the result of a toxic action but simply the densification of heartwood caused by these deposits. Oak, western red cedar and merbau, for example, contain infiltrates which react with iron, causing corrosion of iron fixings and associated staining of the wood. Other forms of unexpected darkening or staining patterns can often be attributed to fungal or bacterial action, sometimes resulting in an unexpected decorative value, despite loss in strength. In some other cases there are perfectly normal explanations for unexpected colours, such as the red colouration to compression wood on the underside of spruce and silver fir branches. Simple chemical treatments can be used to modify colours; lime and hydrogen peroxide are frequently used to lighten wood whilst ammonia reacts with tannin in oak to cause darkening. Steaming generally darkens wood, as in beech, sycamore and particularly walnut sapwood.

Lustre

Colour and figure are only part of the explanation for the attractiveness of certain woods. Lustre is the ability of cell walls to reflect light and is particularly apparent in East Indian satinwood, lauan and sapele. Other species such as hornbeam are very dull and lifeless in comparison. Generally quarter sawn surfaces give the best lustre which greatly enhances figure. However, almost all decorative wood is finished in some way and the ability to take polish, varnish or lacquer may be even more important than the natural lustre of the wood itself.

Odour

Woods generally have a characteristic odour, very apparent when freshly cut, which is associated with their infiltrates. In some cases odour can assist in the rapid identification of woods of similar appearance, such as the separation of true cedar from pencil cedar (Virginian juniper) and cigar box cedar *(Cedrela odorata)*. Virginian juniper is used for pencils because of its even texture and excellent workability but *Cedrela odorata,* which is superficially similiar to mahoganies and only distinguishable by odour, is used for cigar boxes as it is said that its spicy aroma improves the quality of tobacco. The odours of some woods make them unacceptable for certain purposes. Some wood odours contaminate foodstuffs whilst others are toxic to plants or even corrosive; the vinegar smell of some conifers, often masked by the resinous odour, is an indication of the liberation of acetic acid. Many odours are pleasant, such as fresh cut oak, camphor wood and cedar, but others are distinctly unpleasant such as the foetid smell of Australian walnut *(Endiandra)*. The odours of jam wood, sneeze wood and camphor wood are self-explanatory; Formosan camphor wood is often used for storage boxes as the odour repels insects. Some Australian *Acacia* spp. smell of violets, coachwood is reminiscent of new-mown hay and West Indian satinwood has the odour of coconut oil.

Irritants

Most people involved in the woodworking industries are already aware of a few potentially irritant woods. The dust arising during sawing or sanding generally affects the respiratory tract, the eyes or the skin, resulting in coughing, sneezing, breathing difficulty, streaming eyes, nose-bleeding and skin irritations of varying intensity. In some cases dust can cause general disturbance leading to headache, nausea and vomiting, giddiness and vision disturbance. Dermatitis can vary in different sufferers; indeed individual sensitivity is perhaps the most significant variable factor. Mild cases show slight reddening of skin accompanied by itching. In more severe cases a burning sensation will be accompanied by development of a rash, ultimately leading in extreme cases to blistering and weeping sores, perhaps largely as a result of scratching. All symptoms gradually subside when the source of irritation is removed. However, dermatitis can be simple primary irritation or can lead to sensitization. In a case of primary irritation contact with the irritant will result in the symptoms previously described, which will disappear with the removal of the irritant. Subsequent re-exposure to the irritant will lead to the same symptoms developing to a similar degree. If sensitization occurs the initial symptoms are the same but the body becomes sensitized to the irritant and re-exposure results in much more acute symptoms, even if the irritant intensity is low. All these symptoms result from particular infiltrates or extractives, specific chemical compounds within particular wood species, rather than wood particles as dust. Allergic individuals therefore develop dermatitis only when in contact with dust of a species of wood containing a particular irritant chemical but it must be emphasised that, whilst troublesome woods are usually known to all those working in the industry, they normally cause nuisance rather than a health hazard. Severe symptoms are extremely rare and the incidence of wood irritancy is low compared with other industrial ailments.

The Princes Risborough Laboratory has listed the irritant wood species that are most commonly reported in England. Guarea *(Guarea* spp.), mansonia *(Mansonia altissima),* western red cedar *(Thuja plicata)* and white peroba *(Paratecoma peroba)* are most frequently reported as causing both respiratory and dermatitic irritation. Afrormosia *(Pericopsis elata)* and peroba rosa *(Aspidosperma* spp.) are less often reported as causing both symptoms. Dahoma *(Piptadeniastrum africanum)* is frequently reported as a cause of respiratory irritation alone whilst abura *(Mitragyna ciliata)* and makore *(Tieghemella heckelii)* are less frequently involved. Iroko *(Chlorophora excelsa)* is frequently reported as a cause of dermatitis whilst agba *(Gossweilerodendron balsamiferum),* keruing *(Dipterocarpus* spp.), rosewood *(Dalbergia* spp.) and teak *(Tectona grandis)* are less frequently reported. A particularly interesting case is African mahogany which is sometimes reported as causing dermatitis but this is attributed to *Khaya anthotheca* alone; in Britain *K. ivorensis* is most commonly used whilst *K. grandifoliola* and *K. anthotheca* have only been imported in significant amounts during the last few years, so that dermatitis associated with African mahogany is a fairly recent occurrence. In fact, it is normally more common in France which imports mahogany from the Ivory Coast and the Cameroons where *K. anthotheca* is the dominant species.

Density

The density or weight per unit volume of a piece of wood is a particularly important property. Density is most conveniently expressed as specific gravity, or the density relative to the density of water. Wood substance has a specific gravity of about 1.55 but wood itself consists of a mixture of wood substance and spaces, the amount of wood substance per unit volume deciding the dry density which can vary in common species from about 0.3 to 0.8. If the spaces are filled with water this greatly increases the density, so that green or living wood which normally has a moisture content of between 60 and 200% may have a density in excess of 1.00 so that logs cannot be floated as a means of extraction from the forest unless they have been pre-dried in some way. Teak trees are often ringed two or three years before felling by cutting through the bark and sapwood in order to kill the tree and allow it to dry out. In fact, most of the drying occurs shortly after ringing as the leaves remove most of the remaining water from the sapwood. In northern temperate forests trees are traditionally felled during the frozen winter months when the moisture content in the living tree is at its lowest, although the excessive demand for wood in recent years has tended to result in felling throughout the year and the use of road transport rather than the traditional floating on rivers flooded by the melting snow.

Obviously density affects the cost of transporting wood by road, rail or ship but this is a matter of ensuring that the moisture content is as low as possible. Moisture content is also important where wood is used as a packaging, such as in cargo containers, boxes and crates, but there are other more sophisticated uses where the maximum strength to weight ratio is essential, as in the manufacture of ladders, racquets and bats. In fact, a high strength to weight ratio is one of the principal advantages of wood compared with competitive materials, although sometimes cheapness, ease of working or low thermal conductivity may be the reason why wood has been selected for a particular use. For some purposes the strength to weight ratio is not so important and high density is preferable, as in the manufacture of furniture, weaving shuttles, bowling woods and croquet balls.

Wood is considered to have moderate density if its dry specific gravity lies between about 0.36 and 0.50, so that woods below this range are light density and those above are heavy. It should be noted that, in many parts of the world, density is not expressed in terms of oven dry specific gravity but in Imperial units; thus European beech has a density of about 43 lb/ft^3. In theory these Imperial units can be converted to specific gravity by dividing by 62.4, the density of water in lb/ft^3, but the conversion is only approximate as Imperial units normally refer to densities at a "normal" moisture content of 12%, a standard moisture content that is often applied when measuring the physical properties of wood.

Hardness and toughness

Wood is frequently described in terms of its hardness and toughness but these are terms that are difficult to define. Sometimes wood is said to be very tough because it is difficult to saw or plane, or it has good resistance to abrasion, splitting or shock loads. In fact, the ability to resist excessive shock is probably the property that can best be described as toughness. The meaning of hardness is perhaps even more obscure; the wood substance has a hardness of about 2.5 on the geologists' scale starting at 1.0 for talc, which can be scratched with a fingernail, and which finishes at 10.0 for diamond. This assessment of the hardness of wood is entirely meaningless as few woods even approximate to solid wood substance and the hardness of a piece of wood really depends upon the proportion of wood substance to space, although it is considerably influenced by the orientation of the wood substance, so there is no simple relationship between density and hardness. Perhaps the simplest definition of hardness is the ability of wood to resist penetration, usually measured by the force required for a standard ball to penetrate for half its diameter. Hardness decreases with increasing moisture content so that it is necessary to standardise the moisture content, perhaps at 12%, when carrying out tests to compare various samples. Hardness also varies substantially depending upon the orientation within a piece of wood, so that the hardness or ability to resist penetration is far higher for an end grain surface than for a radial or tangential surface.

Occasionally hardness or toughness is used to describe the ability of a wood to resist wear or abrasion, so that hard or tough woods such as hornbeam or apple were once used for silent cogs, and maple is used for dance floors. In fact, hardness is not the only property involved and resilience to accept and later recover from an excessive point load may be rather more important than the ability to resist simple penetration. Another aspect is the variability within a wood sample; in a scuffing test to represent walking a flat surface might be dragged across a piece of wood and would indicate the average properties of the sample, whereas a simple abrasion test involving sand blasting might show, for example, that the spring wood in softwoods has far less resistance than the summer wood.

Specific heat

The thermal properties of wood are extremely important. Specific heat is the term used to describe the amount of heat energy that is required to raise unit mass of the material through one degree of temperature. The specific heat of wood is comparatively high, four times as high as that of copper, but this relates to the mass of material. The density or specific gravity of wood is very low so that the specific heat per unit volume is also very low compared with competitive materials such as metals and concrete. However, the specific heat of wood is of minor importance compared with its thermal conductivity, or k value.

Thermal conductivity

The thermal conductivity of wood is approximately 0.4% of that of steel and 0.05% of that of copper, having the same order of value as cork and gypsum plaster which are commonly used for insulation purposes; indeed balsa is similarly used. The thermal conductivity varies approximately in proportion to density, being lowest for low density wood such as spruce and highest for high density woods such as rock elm. The thermal conductivity also increases significantly with moisture content, by perhaps a third for a moisture content increase of 40%. However, generally the low specific heat per unit volume and the low conductivity combine to ensure that wood feels warm to the touch and, in structural applications, wood claddings or floors provide excellent thermal insulation.

Moisture content

The properties of wood are profoundly influenced by the presence of water. "Green" wood, i.e. the standing tree or freshly felled wood, has a very high moisture content, varying typically from 60–200%. The green moisture content tends to vary inversely with density, so that black iron wood with a specific gravity of 1.08 has a green moisture content of only 26%, whereas South American balsa with a specific gravity of 0.2 has a green moisture content of about 400%. During drying the "free" moisture is first lost from the cell spaces and this involves little change in properties except for change in density. Eventually drying will result in the wood reaching the fibre

Fibre saturation point

saturation point, normally at a moisture content of 25–30%, where further drying results in the loss of "bound" moisture from the cell walls. The loss of bound water typically results in shrinkage and a progressive change in physical properties. In addition the proportion of bound water remaining in the wood is approximately proportional to the relative humidity of the surrounding

Hygroscopicity

atmosphere; wood is hygroscopic and will gain or lose water with changes in atmospheric humidity, and these changes will result in swelling or shrinkage In fact, the moisture content of the wood tends to lag behind the changes in the relative humidity of the atmosphere, a phenomenon known as hysteresis. Most of the changes in properties with variation in moisture content can be attributed largely to the submicroscopic or chemical structure of wood and are discussed more fully in Appendix 2, whereas the measurement of moisture is covered in Chapter 2 and methods for preventing changes in moisture content in Chapter 3.

Movement

The swelling or shrinkage with changes in moisture content is known as movement. In the longitudinal direction the movement is generally very low, only about 0.1% for a change of moisture content from normal "dry" wood at about 12% to the "wet" condition at fibre saturation point or above. In contrast the movement in the radial direction for the same moisture content change is about 3–5%, and in the tangential direction 5–15%. The difference in movement between the radial and tangential direction can be largely attributed to the fact that spring wood shrinks less, so that in the radial direction the movement can be attributed to the average of the spring and summer wood shrinkage whereas in the tangential direction the dense and strong ribs of late wood tend to control the physical behaviour and thus generate higher movement. Indeed, if a piece of softwood is planed smooth at a low moisture content and then wetted, the summer wood will expand to a greater extent than the spring wood so that a corrugated surface will be produced. Generally a higher lignin content in a particular wood species will result in lower movement, or greater stability. The high shrinkage in the tangential direction, coupled in some species with weakness induced by large medullary rays, sometimes results in pronounced surface splits developing during drying. However, it is explained in Appendix 2 that the movement characteristics can be largely attributed to the microfibril angle in the cells, and as this changes progressively from the pith outwards, there is also a progressive change in movement. Heartwood close to the pith is most stable but, in addition to this progressive change, there is a far larger alteration in movement between heartwood and sapwood in many species, the sapwood often possessing extremely high movement. These changes in movement relative to the original position in the tree, coupled with distorted annual rings and twisted grain, can result in warping or splitting, but these defects will be considered in detail later in this chapter.

Stability

Water vapour is lost or gained most rapidly through end grain surfaces

Figure 1.4. **Shrinkage on drying.**

because of their high permeability, a factor that also encourages the very rapid absorption of liquid water. Whilst this means that many defects arising through moisture content changes are concentrated around end grain surfaces, such as splits and the development of decaying or staining fungi, in general wood is used as long pieces with only a limited end grain surface so that the side grain is probably more significant, particularly in service in dry conditions where it is the hygroscopicity of the wood coupled with changes in atmospheric humidity that are likely to have the greatest effect. In principle wood softens as moisture content increases towards the fibre saturation point, a property that may be very valuable in some circumstances, such as when bending wood or peeling veneers from logs. However, this softening substantially affects strength so that, in structural uses, wood must be employed in adequate dimensions to tolerate the strength loss that may occur should the moisture content approach the fibre saturation point.

When trees are first converted to sawn wood the moisture content is significant because of the danger of fungal damage if a high moisture content is maintained, and also because of the excessive transport costs that are involved if the weight of wood is perhaps doubled by an excessive moisture content. However, if green wood is used in buildings the most serious problem, other than the danger of fungal decay, is the cross grain shrinkage that will occur as the wood gradually achieves its equilibrium moisture content with its surroundings. This is a very serious problem in certain uses, such as window and door frames, floor boarding, panelling and the many other uses where cross grain shrinkage will result in the development of unacceptable gaps between the individual pieces of wood. It is therefore usual to air season or kiln dry wood before installation to the average moisture content that it will encounter in service, about 12–15% in freely ventilated parts of the structure, 8% in normal centrally heated buildings but perhaps 4% in centrally heated buildings in very cold climates where the winter relative humidity is exceptionally low. The main problem is that wood may be processed to the required moisture content but this may change during storage, delivery or installation, particularly in buildings where the "wet" trades such as bricklaying and plastering result in very high humidities during construction. Indeed floors can sometimes be laid very promptly after kiln drying in a recently plastered house where the high humidity will cause them to expand, sometimes the entire floor lifting in a dome, or the expansion causing serious damage to the restraining walls. Stacking the wood for a period inside the building before laying will avoid this particular problem as it will reach equilibrium with the high humidity in the damp new building but, when the building is ultimately completed, the humidity will fall. Quarter sawn floorboards will simply shrink in width, leaving gaps at the joints, but flat sawn boards will cup in addition; they should be laid with the heartwood surface upwards so that each board is raised in its centre rather than at the edges where it may cause tripping. The integrity of the floor and thus its ability to act as a firebreak or thermal insulator is naturally affected by this shrinkage, unless the boards are joined by tongues to span the shrinkage gaps.

Plywood

A floor is, of course, a rather large panel, normally assembled out of separate boards, and the problems encountered in floors are naturally encountered in other panels such as wall linings, door panels and furniture. A number of panel materials are now available in which wood has been processed to minimise these shrinkage defects. In plywood veneers of wood are orientated at right angles so that the stable longitudinal grain direction is used to physically restrain the movement in the cross grain direction. This method for reducing shrinkage is particularly successful in normal buildings but sometimes fails under extreme conditions, as when used in boat building. Adhesives have been developed which will withstand the very high stress that is developed when the moisture content fluctuates from fibre saturation point down to quite low humidities but if this occurs regularly there is a tendency for the wood to rupture on either side of the glue line and this can only be avoided by using wood of moderate or low movement in the manufacture of the plywood. It should be appreciated that the plywood still expands normally in thickness but this is seldom a problem, unless it affects the weather seal when plywood is used as exterior panelling in buildings or boats.

Particle board

One particular advantage of plywood is that, as the grain runs longitudinally in both directions within the plane of the board, they are equally strong. In contrast the alternative process for producing a panel, the manufacture of particle board, results in a material that is equally weak in both directions. In particle board manufacture the principle is to divide the wood into small particles which are then randomly orientated so that the movement is shared equally in all directions, although it is restrained to some extent if sufficient adhesive is used. The movement is higher than for plywood and, whereas plywood presents a completely smooth surface, a particle board is comparatively irregular, a defect that is particularly noticeable with changes in humidity as the surface chips expand and contract across their grain. In attempts to minimise this defect smaller particles are sometimes used for the surface or the board is veneered. Unfortunately veneers are often applied on boards with comparatively large surface chips in which any movement results in deflection in the veneer or "shadowing".

Fibre board

The only alternative is to reduce the wood to individual fibres which are then randomly orientated and reconstructed into a board which is, in effect, a rather thick sheet of paper. Low density panels are known as insulation board and are used for lining walls and ceilings, but the medium and high density panels, or hardboards, present a reasonably durable and sound surface for painting. Whilst fibre boards might appear to be similar to plywood in principle, although the wood components are re-orientated on rather a different scale, they do not achieve the same end result. In fibre boards the movement tends to be randomised in all directions, whereas in plywood the movement in the plane of the board is physically restrained. Plywood is therefore preferred for external panelling and other situations where it may achieve a high moisture content as its movement within the panel dimensions remains small, whereas a fibre board will swell significantly in similar conditions and this may lead to serious buckling.

Strength

There are a number of factors that affect the strength of normal wood. In general the strength is approximately proportional to the density. However, fibre length is important, particularly in relation to bending strengths. Woods that suffer from a short growing season, such as conifers growing close to the Polar regions, are often dense but their fibre length is comparatively short and their bending strength suffers accordingly. Whilst they should not therefore be used for high stress applications, such as beams and joists, their close rings and high density usually give excellent working properties. Strength is reasonably consistent within an individual tree, and sound sapwood is normally as strong as sound heartwood. Wide rings lead to lower bending strength when loaded on a tangential surface, apparently because the spring wood compresses, but this may be coupled with increased toughness, or ability to withstand shock loads. Because of the tendency of spring wood to compress, a beam is generally stronger if loaded on the quarter sawn face. A high degree of lignification, as in tropical hardwoods, usually indicates high compressive strength though low toughness; the lignin deposit leads to brittleness and low flexibility.

Elastic limit

If wood is subjected to load which is subsequently removed it will return to its original state, provided the load was not in excess of the elastic limit, or the limit of proportionality; up to this limit the deflection is proportional to the load but beyond this point the deflection becomes excessive, leaving a permanent set or distortion when the load is removed. The rate of loading is not important within the elastic limit but the period of loading can be very significant; creep or fatigue will occur which can reduce the bending strength to only one half or three quarters of the normal when, for example, a beam is under load for a protracted period such as fifty or one hundred years. Increasing temperature leads to a decrease in strength, although only to a limited extent under normal temperature conditions; moisture content is far more significant.

Figure 1.5 **Limit of proportionality.**

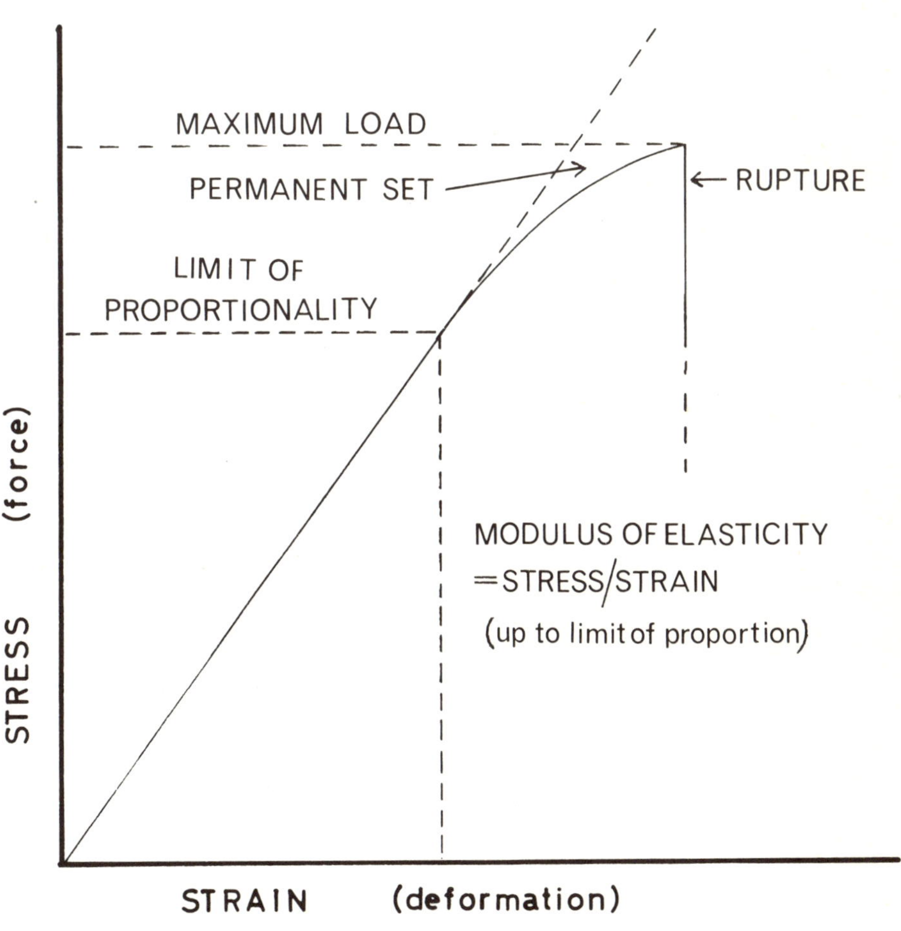

Strength and moisture content

Moisture contents in excess of fibre saturation point have no influence on the strength of wood. Dry wood tends to be more brittle and the strength properties generally increase with decreasing moisture content, except for toughness which decreases with moisture content in most species; Sitka spruce, for example, is an exception as toughness increases with decreasing moisture content. The changes in strength with moisture content are very significant. If normal dry wood at about 12% moisture content is wetted to fibre saturation point it will lose about half its bending strength and compressive strength along the grain. Incidentally the loss in compressive strength with increasing moisture content will be less than the loss in tensile strength, a particularly important factor in wood bending where steaming is frequently used to plasticise the wood by increasing both its moisture content and temperature. Another interesting observation is that wood that is dried, re-wetted and re-dried to the same low moisture content is both more brittle and more stable than wood that has dried in a single stage.

Figure 1.6 **Variation of strength with moisture content.**

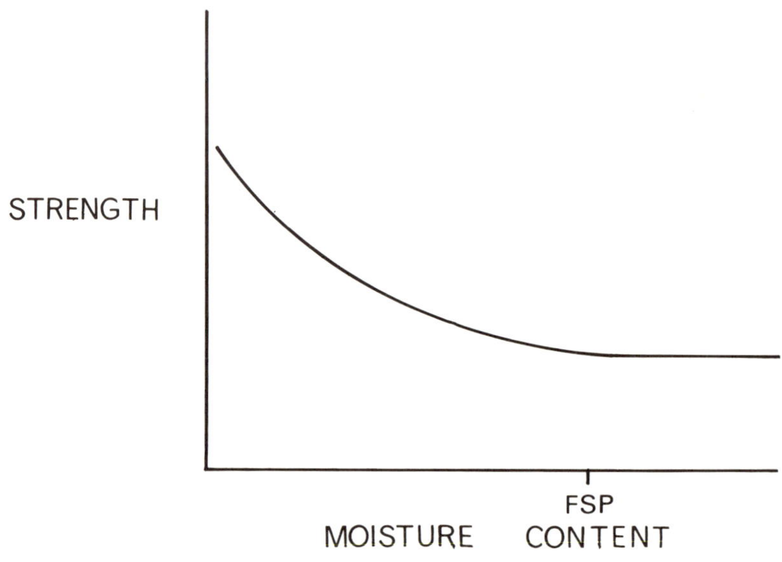

Reaction wood

Irregular grain, splits and checks, fungal decay and abnormal microscopic structure all significantly affect the strength properties of wood. Reaction wood is generated under constant but abnormal loads, such as in branches or sloping trunks where the upper tissue is in tension and the lower in compression, or in trunks subject to a relatively constant wind load. In softwoods the reaction to these constant loads is the formation of "compression" wood within the compressed tissue, whereas in hardwoods the reaction is the formation of "tension" wood within the tensed tissue, the upper side of branches or leaning trunks. Compression wood in softwoods usually has wide growth rings, apparently with a high proportion of summer wood. This unusually high proportion of red coloured summer wood explains the term "rotholz" which is used in German to describe compression wood in softwoods. The darker and redder colour of compression wood is an indication of the higher lignin content, perhaps increased from the normal level of about 20% to 30% and accompanied by a corresponding reduction in the cellulose content of the cell walls. This higher lignin content explains the brittleness of compression wood, its high compressive strength and low hardness compared with normal wood. In addition compression wood tends to shrink significantly in the longitudinal direction on drying and, where compression wood forms the edge of a piece of normal wood, this often results in the development of cross grain checks which substantially reduce the strength.

Tension wood in hardwoods is usually associated with fibres tearing loose during sawing, resulting in a rather woolly surface and choking of the saw teeth. The detachment of fibres in this way also interferes with veneer cutting. Tension wood tends to be higher in cellulose and lower in lignin than normal wood and this leads to loss of hardness and excessive shrinkage, although these defects are not serious as with compression wood in softwoods.

Strength and grain

Sloping grain can lead to very significant loss of strength. Diagonal grain is the term usually used to describe sloping grain arising through careless conversion which results in the grain or annual rings having a distinct angle to the surface of the piece of wood. Spiral grain can arise, however, in a correctly cut piece of wood where the tracheids or fibres in the tree tend to have an angle to the vertical and thus follow a spiral pattern within the annual rings. Sloping grain is very serious in, for example, chair legs and beams where it can lead to substantial loss in strength, as previously explained. Even wavy grain can reduce compressive strength by as much as 30% or 40% so that grain orientation is a very important factor affecting the strength of wood. Warping, or the bending of the growth rings or grain within the wood, can also lead to substantial loss in strength and a bend of only 1 in 1,000 in a long column may lead to a strength reduction of as much as 20%.

Strength and knots

The influence of splits and checks depends largely upon their extent and their position within the piece of wood but knots may be even more significant. There are two ways in which the influence of a knot may be represented. One way is to consider that the knot represents a space and thus loss of strength. The alternative is to observe that the grain and annual rings are distorted around the knot, so that strength is reduced by the changing orientation of the grain. This results in the grain emerging from the edge of the wood so that the loss of area theory of loss of strength may still be appropriate, particularly with regard to compressive and bending strength, although knots do not reduce stiffness. Obviously knots are most significant when they occur in positions of maximum tensile stress, such as at the bottom of beams, and they are least significant when in compression, as at the top of beams.

Knots

There are various descriptions for different types of knots. A typical knot arises where the branch emerges from the trunk. The branch wood is in most respects similar to trunk wood and the only influence of the knot is the orientation of the grain. Where a knot arises from cutting through a living branch it is known as a live or tight knot and is only significant in the mechanical sense through the deflection of the grain. However, branches are often removed through accidental damage to the living tree, through death from mutual shading and through deliberate trimming. These dead stumps are eventually overgrown and concealed but their significance depends upon when they occurred in the life of the tree and whether the branch broke off flush with the trunk or left a dead stump projecting. Lower branches are generally lost when the tree is comparatively small. The resultant knots are therefore confined to the inner wood of the trunk and the outer wood remains free from knots. If a branch is removed flush with the trunk the stump will rapidly be concealed and the outer parts of the tree will develop normally. A board cut from the trunk will disclose the knot but it will be tight. If the stump projects it will take many years to overgrow and there will be a tendency for the knot to be surrounded by old bark or cork, giving a dead or loose knot.

Knots seriously affect strength. In stress grading where wood is selected

Stress grading and knot types

either visually or by machine to ensure consistent strength properties, knots will generally result in reduced strength and thus lower grading. Knots are often classified as small pin knots, resulting from the loss of early branches, spike or splay knots where the stump is cut lengthwise, as in quarter sawn boards, and encased knots which are dead knots which are not easily dislodged and which are classified as perhaps small, medium or large.

In conifers a tree will develop a leading shoot each year and the limit of this leading shoot will decide the position for the growth for the following year. The growth will radiate from this position, giving a further leading shoot and also a surrounding ring of branches, known as a whorl. These branches will naturally result in a ring of knots and thus a weak ring within the wood which, because of its distorted grain, is of particular importance in compression uses, such as pit props. In larger round timber, such as poles, piles and masts, only the inner wood is affected by knots occurring during the early life of the tree and there is a surrounding tube of sound wood free from knots.

Spring

Spring is a defect that is frequently seen in quarter sawn boards where one longitudinal edge has shrunk relative to the other, causing curving within the plane of the board which otherwise remains flat. Spring is apparently the result of tensions within the growing tree which are released upon sawing and most frequently occurs in boards close to the centre of the log. In a flat sawn board through the centre of a log spring may become apparent, particularly in susceptible woods such as elm, as end splits or shakes.

Splits, shakes and checks

Shakes and checks are terms that are frequently used indiscriminately to describe splitting, although it is probably correct to use shakes to describe defects inherent in the growing tree which become apparent on conversion, and checks to describe defects arising through seasoning or kiln drying. There are two principal sources of tensile weakness in wood, the medullary rays in hardwoods and spring wood in both softwoods and ring porous hardwoods. Any tensile or sheer stress will therefore tend to result in splits radially along the medullary rays or tangentially along the spring wood. The addition of a new annual ring under the bark is accompanied by the death of the innermost sapwood ring and its conversion to heartwood. The living cells in sapwood conduct moisture from the roots to the leaves but the conduction paths become obstructed in heartwood which progressively dies over a period of many years. This loss of moisture content can lead to the development of heart shakes, sometimes known as heart checks or rift cracks, or even star shakes or box heart when the cracking gives an appropriate pattern around the pith or centre of the log. Alternatively cup or ring shakes develop where the split occurs tangentially and follows the spring wood.

Frost cracks and ribs

In very cold climates frost cracks can develop in the external surface of the tree. This damage is caused when unusually late or early frosts lead to freezing of the sapwood whilst still at an exceptionally high moisture content. Frost cracks are longitudinal and in the radial plane, and tend to open and close whenever suitable freezing conditions are repeated. As a result a callus is formed over the open end of the split as the tree attempts to heal the damage and this gives rise to a frost rib projecting beyond the normal rings and into the bark. Whilst frost cracks are undoubtedly caused by frost, thunder or lightning shakes are unlikely to be caused by thunderstorms.

Thunder shakes, cross shakes and brittleheart

Thunder shakes are a cross grain defect which severely reduces the strength properties of the wood which therefore becomes almost worthless. The shakes are clearly visible as a line across the grain of the wood and microscopic examination shows that the fibres have been distorted and broken through compression. Thunder shakes, which are sometimes more appropriately called cross shakes or brittle heart, are frequently observed in the softer tropical woods such as African mahogany, lauan and meranti. Thunder shakes are sometimes associated with spongy or punky heart. They are usually invisible in green timber but become very apparent on drying; indeed, a dry board may break when lifted, an observation that has resulted in faulty boards being described as "three-men-boards", the third man supporting the board at the break!

Resin or pitch

There are a number of distinctive defects which can be observed in wood. If the outer cambium layer of softwoods is damaged a protective resin exudate protects the wound before neighbouring tissue eventually grows over it. This included resin gives pitch pockets following the line of the annual rings. These resin deposits are known as pitch blisters when exposed on the tangential plane but pitch seams or veins when seen as a narrow defect on a quarter sawn face. Pitch seams are an open space filled with resin and they must not be confused with resin or pitch streaks which are narrow brown streaks progressively fading out along the grain. This defect, frequently observed in spruce, Douglas fir and other conifers, is caused by local accumulations of resin spreading along the tracheids. It is not readily apparent that the brown colouration is caused by resin and this defect should not therefore be confused with incipient decay; resin streaks do not cause the softening characteristic of decay and do not affect strength. Strawberry mark is a rather similar minor defect caused in Sitka spruce by resin accumulations. It generally appears as a red or brown zone, 1″ (2.5 cm) wide and perhaps $1\frac{1}{2}$″ (4 cm) long on the radial face and a thin bar of colour on the tangential face. Strawberry mark is a form of resin or pitch streak and does not affect strength.

Gum or latex

Resin or pitch deposits are confined to softwoods but similar features are often found in hardwoods. Latex canals occur only in a few woods in, for example, the families Apocynaceae and Sapotaceae. These are regular natural features but they substantially reduce strength where they occur and these woods are therefore weaker than other tropical hardwoods. Gum veins or canals filled with dark deposits frequently occur in some tropical hardwoods. In some species, such as jarrah, they are common but widely scattered so that they have an insignificant effect upon strength. Gum veins are common in wood from fire damaged forests, presumably through damage to the cambium, and can then lead to serious loss of strength. The occurrence of gum veins is a characteristic of some species, such as African walnut, where it contributes to the figure.

Mineral streaks

Mineral streaks is a term used to describe local discolouration, usually dark streaks or patches, which commonly occurs in tropical hardwoods but does not impair strength. Mineral streak discolouration must not be confused with bacterial or fungal staining which often leads to similar darkening but also softening and loss of strength. Certain Dipterocarpaceae such as lauan, meranti, seraya, keruing and gurjun are often said to have mineral streaks but these are actually resin canals in longitudinal section, containing white or yellow resin which contrasts against the red or brown colour of the wood itself.

Included phloem or bark

Occasionally a vertical band of weak wood occurs in the plane of the annual rings in some species. There is often no apparent explanation for this defect but microscopic examination shows that the band of weak wood is histologically similar to phloem and the defect is therefore described as included phloem, although the mechanism of inclusion is far from clear. In some temperate hardwoods such as alder, birch and sycamore insect damage to the outer cambium can cause small cork or bark inclusions to occur within the annual rings. These appear as small brown dots or streaks in the wood, usually described as pith flecks. Occasionally more severe cambium damage will result in the inclusion of larger bark pockets. These often occur in cross section as a false brown ring which is often described as a frost ring as it is caused by late or early frost damage of immature cambium.

Seasoning defects

It has already been explained that wood shrinks when its moisture content is reduced, that the outer zones of the trunk shrink more than the inner and that the sapwood shrinks more than the heartwood. In addition the outer layers of a log or piece of wood are ventilated and thus dry more rapidly than the inner, a combination of all these effects leading to the distortion and tissue rupture defects which are a characteristic of faulty seasoning or kiln drying. One particularly important point is that wood is plastic when wet but set when dry so that it is essential that wet wood should be carefully stacked flat to avoid unnecessary distortions.

Warping

Warping is the general term for seasoning distortion and takes several forms.

Cupping

Cupping is warping across the grain or width of a board and arises in flat sawn boards through the outer zones of the trunk, particularly the sapwood, shrinking to a greater extent than the inner zones, giving a planoconcave surface on the outer side of the board relative to the trunk. Flat sawn boards used for flooring should always be laid with the heart upwards to give a planoconvex surface and thus avoid trip edges. The alternative is to use more expensive quarter sawn boards which are resistant to cupping and also more hard wearing.

Twisting

Twisting arises in boards through the presence of spiral or interlocking grain and is really the result of wood selection rather than a seasoning defect, although spiral grain is quite common and perhaps unrealistic to reject.

Bowing

Bowing is longitudinal curvature arising perpendicular to the surface of a board, either through spring in a flat sawn board but more usually as a result of sagging in the stack during drying. Spring or crooking is longitudinal curvature within the plane of the board and can be severe in some species such as elm or kempas grown in swampy areas, giving decreased strength. Finally diamonding occurs in pieces of wood of rectangular section when the annual rings pass diagonally across the end grain. The radial shrinkage on drying is less than the tangential shrinkage so that the diagonals in, for example, a square section piece of wood become different in length and give a diamond shape in cross-section.

Fig 1.7 Diamonding or unsymetrical shrinkage in the tangential and radial directions.

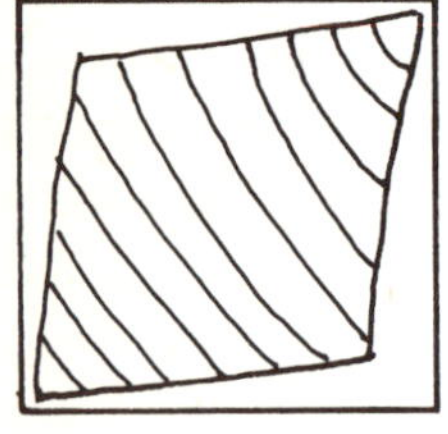

Drying occurs most rapidly at the end grain through its very high permeability. The shrinkage close to the end grain may therefore occur whilst the rest of the piece of wood still has a high moisture content and retains its swollen dimensions. The stress between the shrinking end grain and swollen inner wood frequently results in splits on the end grain surfaces, although splits is a term normally used to describe cracks passing right through the piece of wood.

Drying checks

In contrast checks are more shallow, occurring as end checks on the end grain or surface checks on the other faces, perhaps diagonal to the edge of the board if spiral grain is present. Splits and checks may close with rewetting but, although they may then be virtually invisible, the strength has been lost. Serious splits are sometimes described as shakes, although it has previously been explained that this term is more correctly used to describe defects inherent in a particular tree rather than those developing as a result of faulty seasoning or kiln drying.

Case hardening

The stresses and strains resulting from the rapid drying of end grain in advance of the rest of a piece of wood has already been described, but an identical situation occurs in respect of the entire surface of the piece of wood which naturally dries more rapidly than the inner wood and thus shrinks in advance. This defect is known as case hardening. Obviously the tendency of the outer zones to shrink in advance of the inner results in tensions which can only be relieved by the rupturing of the outer zone or the development of surface checks. If rupture does not occur it is possible for the outer zone to be tension set so that subsequent sawing will release this tension and result in spring or cupping; if tension set is suspected it can be relieved by steaming. The tension in the outer zone may alternatively compress the inner zone which is, of course, wetter and thus more plastic. Further drying may then result in the subsequent shrinkage of the inner zone without associated shrinkage of the dry set or case hardened outer zone, giving internal ruptures known as honeycomb checks or hollow horning.

Tension set

Honeycomb checks

Normal round knots may present an end grain surface within the face of a normal board. These knots behave like any other end grain surface, perhaps checking or shrinking. In the case of encased knots, surrounded by included bark, shrinkage generally results in loosening and perhaps loss of the knots.

Collapse

Some woods, particularly Eucalypts, acquire a corrugated surface when dried too rapidly, usually through the use of a low humidity or excessively high temperature in a kiln or very fast air drying. This defect is known as wash-boarding or collapse, and is caused when the water content is reduced so rapidly that air is unable to replace it within the cell cavities. The removal of the water from the cell cavities results in the collapse of the weaker spring wood and thus the development of the corrugated surface that has been described. This is a particularly serious problem in fast grown Tasmanian oak, actually a Eucalypt, and also in western red cedar. When collapse is observed the wood can be reconditioned by steaming, the high moisture content and high temperature plasticising the cell walls and permitting them to recover their original size and shape.

Wane

There is one very simple defect that has been omitted until now from this brief description. Naturally the external surface of the trunk of a tree is curved and covered by bark so that the conversion of the trunk into rectangular sections will occasionally result in pieces which will run out through the bark, having chamfered edges, perhaps with bark adhering. These lost edges are known as wane, whether or not the bark is still present, and represent a significant loss in cross-section. Naturally wane occurs at the edges which are under the maximum stress when loaded in service so that the loss of cross-sectional area has occurred at the most critical point. In addition to this simple loss of strength, the presence of wane indicates the presence of outer sapwood which is particularly susceptible to certain wood boring insects, as described in greater detail in Chapter 3.

Fire resistance

Although wood is combustible and used as a fuel, it has surprisingly good resistance to fire. Firstly, wood will only ignite when a temperature of about 270° has been exceeded. Even then the burning is confined to the outer surface, partly because the wood is compartmented so that necessary oxygen cannot easily reach the deeper wood but also because the low thermal conductivity and high specific heat tends to ensure that the inner wood remains at a comparatively low temperature. Charcoal tends to form on the surface of the wood, further restricting the burning rate so that the loss of strength, which depends upon the rate of formation of charcoal, is comparatively slow. In addition the wood remains intact and comparatively stable, even when burning fiercely, thus preserving the integrity of the structure. In comparison metals will expand when heated and lose their strength when a critical temperature is exceeded, whilst concrete structures generally spall violently in fire. Wood therefore has the considerable advantage that it will remain intact in fire and, provided there are no open joints, a wood structure has excellent fire resistance. The main problem with wood is the surface flaming which may cause a fire to spread to a dangerous extent within a building. When used intelligently, by designing buildings to avoid open joints where fire resistance is required and by specifying fire retardant treatments to limit spread of flame, wood possesses unique advantages in respect of fire in buildings.

1.2 Wood properties related to use

The properties of wood in service will depend upon the species, its selection or grading and the manner in which it is utilised. Strength is the most important factor in many structural uses but it requires far closer definition when considering the suitability of a wood for a particular purpose. For example, in a beam the most important factor may be the load carrying capacity, or the load that it can withstand before failure occurs. However, loading may produce a very large deflection which will be unacceptable for many purposes, such as in joists supporting a plaster ceiling where deflection will result in cracking, so that stiffness as well as load carrying capacity is required. In other applications such as tool handles, bats and racquets, stiffness is a distinct disadvantage. Compressive strength along the grain is most important in other applications, such as piles, posts and pit-props, resilience being particularly necessary in the latter case. In ladders the strings or sides must be strong and resilient, properties that are usually achieved by the selection of suitable woods such as Douglas fir, yet they must also be light in weight so that only the minimum dimensions can be employed, a requirement that is usually achieved by careful selection to ensure straight grain and freedom from knots or other defects so that the minimum cross-section can be employed. In contrast the rounds or rungs of the ladder must be strong but stiff and, whereas softwoods are invariably used for ladder sides, hardwoods such as oak with their higher lignin content and greater stiffness are normally used for the rounds.

Safety factors

In ladder strings it is normal to carefully select the components to ensure consistent high strength and reliability but this is not normally possible in general purpose construction wood. Instead it is necessary to incorporate various safety factors in design which will allow for the normal variability in commercial wood. The normal practice is to test small samples of wood in the laboratory but the strength of large structural members may be only 75% of the laboratory calculated values, due to natural variability in grain direction, fibre or tracheid length, and the presence of some knots or drying checks. It is necessary to allow for the creep or fatigue factor when wood is subjected to continuous load and this may, in effect, reduce the strength to about 50% of the calculated laboratory value. Some allowance must also be made for accidental over-loads which occur, for example, during building when construction materials are temporarily supported on a floor or through a building changing in use, a factor that might effectively reduce the strength to perhaps 67% of normal. Finally variability in the selection or grading processes, or in the properties of wood from different sources, reduces the effective strength to perhaps 75% of normal. The combination of these four factors suggests that wood should be used at dimensions which give a safety factor of about 5.5 times the laboratory test results for small, clear and uniform samples. In fact safety factors of between 5 and 7 are required for wood, compared with 3 or 4 for the far more uniform metals such as iron and steel. These safety factors are only concerned with dead loads; live loads can impose a high degree of over-loading which may stress individual components beyond their elastic limit and it is therefore necessary to incorporate safety factors up to 20 in extreme cases such as bridges.

It may thus appear that natural variability limits the usefulness of wood as the safety factors must be rather higher than for uniform structural materials such as steel, yet the exceptionally high strength to weight ratio for wood in general and the stiffness or alternatively the resilience of certain woods ensures that suitable material is available which is not simply adequate but also light, easily worked and eminently suitable for most structural purposes.

Jointing

One of the problems is always the jointing of wood, a fairly simple process in post and beam construction where fixings are only required to maintain the structural elements in position and the load is taken directly on the tops of the posts and the side bearing surfaces of the beams. Whilst wood has very high compressive strength as a post and bending strength as a beam, its excellent strength in tension is difficult to develop because of the need to make a connection between the structural elements. A simple nail, bolt or screw is inadequate when wood is used as an engineering material as it will tend to pull in the direction of the grain. Various connectors have been devised to spread the load, varying from gang nails to spike connectors which are clamped between the two elements by means of a bolt. However, in more sophisticated engineering applications adhesives perhaps represent the extreme development; if the contacting surfaces are correctly prepared the load is spread over a wide surface area and the connection is then vastly more efficient than, for example, welded joints in steelwork where the connection is only a thin ribbon around the jointed area.

Obviously the grain of the wood ensures that compressive strength is greatest in the longitudinal direction, as in the loading of a post, whilst bending strength is greatest across the grain, as in a beam, but there are a number of minor factors that must be considered in addition. Sloping grain will substantially reduce both the compressive strength of a post and the bending strength of a beam, as explained earlier in this chapter, but even the

Ring orientation

orientation of the annual rings in the cross grain direction is also of importance. A beam standing on a radial or quarter sawn face is far more rigid than one standing on a tangential face, the stiffness on the radial face resulting from the vertical planes of comparatively rigid summer wood, and the bending on the tangential face resulting from the compression of the spring wood permitting flexing of the summer wood in the horizontal plane. In a similar way the rigid summer wood of the rings gives quarter sawn wood greater wear resistance when used as flooring, although quarter sawn surfaces are usually preferred because of their freedom from cupping, as ex-

Movement defects

plained earlier. In fact, movement factors such as freedom from cupping are often largely responsible for the selection of a wood for a particular purpose, either to avoid shrinkage gaps or distortion when the wood is exposed to wetting and drying from weathering or changes in atmospheric relative humidity in buildings. Thus flooring and external joinery (millwork) are applications in which seasonal moisture content changes may occur in the wood which must therefore be selected to ensure comparatively low movement. Flooring represents one of the most critical uses for wood as it is necessary to select for adequate strength, low movement and resistance to wear. It is desirable to have resilience or toughness, the ability to withstand impact or high local loads such as stiletto heels. In addition resilience ensures a quiet floor which is less tiring as it avoids the jarring that occurs when walking on non-resilient floors of, for example, concrete. The ageing of wood and ease of maintenance, perhaps largely dependent upon the ability of the floor to accept a finish and maintenance polish, are obviously extremely important, yet often flooring is selected for its appearance alone in much the same way as wood for furniture. Obviously colour, texture and figure are of importance and are indeed perhaps the main advantages of wood over competitive materials in many applications. It is the natural variation in wood that makes it interesting and which, in common with certain stones, so strongly illustrates the attractiveness of natural materials compared with the uniformity of synthetic or manufactured materials.

Appearance

There are a number of other properties of wood that should be mentioned in passing. If wood is exposed to weather the alternating wetting and drying of the external surface will cause stresses which may result in surface checking in woods with comparatively high movement. In extreme cases this checking can lead to severe structural weakening, such as in poles or railway sleepers (ties) used in the tropics. Another danger, for example in preserved transmission poles, is that the checks will penetrate through the preservative treated outer layer which is often confined to the sapwood, thus enabling water to accumulate in cracks penetrating through to the untreated heartwood. In most species the heartwood is darker in colour and this is usually related to the presence of deposits which give it greater durability, or natural resistance to decay, than the sapwood. However, the exclusion of less durable sapwood is often unrealistic and in most applications where exceptional durability is required it is more realistic to employ preservative treated wood. Thus the wood might be selected for its natural durability or alternatively selected because of the ease with which it can be preservative treated; many species have heartwood and sapwood resistant to even pressure impregnation with preservative. In some cases the ability to resist chemical corrosion is important, as in the construction of vats or in buildings subject to corrosive vapours. Usually a wood can be found which will withstand almost any condition, such as abura and southern cypress which both have exceptional resistance to acids.

Durability

Workability and availability

It is little use selecting a wood for its properties related to its end use if it presents particular problems in working. In large scale manufacturing operations the resistance to cutting and the blunting of woodworking machinery is at least as important as the many other selection factors that have already been mentioned. In the case of joinery (millwork) and furniture production the ability to take a smooth surface is also of extreme importance. Finally the selected wood must be available in adequate quantities with consistent properties at a suitable price; slight changes in costs can lead to the rejection of a particular wood and the adoption of an alternative that may be unsuitable, although in many companies it is generally accepted that only large price increases or serious supply difficulties will justify a change which may result in both manufacturing difficulties and perhaps an unreliable product.

The importance of careful selection can perhaps be best gauged by reference to a few examples. Britain has an extensive coal-mining industry which requires a very large number of pit-props. These are generally thinnings derived from the home grown forests and it is found that Scots pine, Corsican pine, Douglas fir, European larch and Sitka spruce all have similar compressive strengths when used as props. The home grown supplies are supplemented with imported props, such as Scots pine and Norway spruce with compressive strengths similar to the home grown material. Maritime pine is also available, principally from Spain, Portugal and Southern France, but its compressive strength is lower than the other woods and it must be used with considerable care.

It is interesting to consider the way in which the properties of some West African timbers are related to their established uses. For example abura and obeche are both used for pattern making where their straight grain, fine texture and small movement are particular advantages. African mahogany, agba, iroko and makore are used for ships' decks and exterior joinery (millwork) where their small movement and durability are important, whilst window and door sills in external joinery tend to be made from woods combining these properties with the ability to machine readily to the special shape of the sill. Indeed the workability is very important, for example for mouldings where fine or medium texture is required, together with straight grain. Turnery, or the use of lathes for the manufacture of furniture legs, tool handles and similar articles, is only possible if the wood is uniform and cohesive so that it will cut evenly in all directions without tearing, although in the case of tool handles it is also particularly important that the wood should be resilient to absorb shock. Thus danta and makore are perhaps the most popular tropical woods for this purpose, although the temperate hardwoods ash and hickory are perhaps more widely used.

1.3 Wood resources

Forest zones

Forest production depends upon geography, climate and other conditions such as soil composition. However the production forests can be divided into major types according to the latitude zones. In the Polar regions the soil is permanently frozen and is unable to support most plants. However, the Polar regions are surrounded by the Arctic and Antarctic Circles which will support slow-growing stunted trees which mark the tree line. The next zones towards the Equator are the coniferous forests with their pines, firs, larches and spruces. These conifers are characterised by their fastigiate manner of growth consisting of the straight trunk with branches radiating in a whorl originating where the shoots developed from the growth limit of the leading shoot of the previous year. Under load the branches and their needles droop, readily shedding their snow load and making the conifers particularly successful in cold climates. These conifers, mostly evergreens, stretch in a band around the world from Scotland through Scandinavia, Russia, Siberia, North China and North Japan to Canada and represent the source of the principal building woods of Europe, North America and many more distant countries; redwood (Scots pine) from Scandinavia and Douglas fir (Columbian pine) from Canada are regularly shipped to Australasia. Naturally the southern hemisphere has a similar coniferous forest zone but it is much less productive in view of its restricted area; the southern hemisphere in these latitudes is dominated by ocean.

Coniferous forests

Figure 1.8 Fastigiate growth in conifers.

The next latitude zone is the temperate forests of Central Europe, North Japan and the north eastern states of the United States of America. The temperate forests produce principally hardwoods, such as birch, poplar and alder, progressing in lower latitudes to oak, ash, beech, chestnut, sycamore, lime, poplar and willow. In North America the principal woods are hickory, various maples and gums, American whitewood (Tulip tree) and magnolia. Most of the temperate forest trees are deciduous, shedding their leaves in winter, but a few are evergreen, such as the holm oak.

Temperate forests

Sub-tropical forests

In the sub-tropical zone an even wider variety of hardwoods occurs. The Mediterranean area supplies box, evergreen and cork oak, chestnut, olive, walnut and plane, as well as some softwood species such as cedars. The sub-tropical forests of the United States, such as Florida and Louisiana, produce persimmon wood, southern yellow pine and the famous Pitch pines of Florida.

Tropical forests

The tropical forests are of particular importance and can be divided into two types. The savannas are areas of low rainfall in, for example, Thailand, India and Africa which consist of open grass-covered plains with only scattered trees. From the wood production point of view the tropical rain forests are far more important and cover vast areas in West, Central and East Africa, South-East Asia, and Central and South America. These are areas of very high rainfall, producing dense forests consisting of a great variety of species, although only a few may have value as the wood of commerce; there are considerable efforts in tropical countries to encourage local utilisation of the "secondary" species of no commercial value. These tropical forests can produce woods possessing an almost infinite variety of properties, from very lightweight species such as obeche from West Africa to the medium weight and fairly hard teak from South-East Asia and the very heavy, strong and durable ekki and afzelia from West Africa. At the present time the most important species are perhaps the *Meliaceia* spp. or mahoganies, including the true mahoganies of Central America such as Honduras mahogany and the virtually unobtainable Cuban mahogany. This family also includes African mahogany, sapele and a large number of closely related species such as guarea and crabwood. Teak is perhaps one of the most valuable tropical woods with its uniformity, stability and durability, but it is only available from Burma and Thailand. Teak substitutes, perhaps a little less stable and a little less durable but highly valuable all the same, are available as afrormosia from West Africa and muninga from East Africa. Greenheart, a strong, heavy and exceptionally durable wood from British Guiana, is widely used throughout the world for marine piles. The tropical forests are also capable of producing many special purpose woods such as the very dense lignum vitae which is used for stern-gland bearings on ships, ebony, rosewood and the exceptionally light balsa, frequently used as an insulation material. At the same time the tropical forests can provide many more general purpose woods and South-East Asia alone provides gurgun, yang, lauan, meranti and seraya, as well as keruing which is widely used as a hardwearing decking for jetties and warehouse floors.

Southern hemisphere

The southern hemisphere naturally mirrors the northern hemisphere in its forest zone pattern, except that the large areas of ocean restrict the coniferous and temperate zones. Softwoods are comparatively rare, largely as a result of excessive felling, but hoop pine occurs in Australasia as well as kauri pine, although the latter species is now virtually unobtainable except in Fiji. Attempts are being made to replant forest areas, such as the *Pinus radiata* plantings in New Zealand. Mention should also be made of the *Podocarpus* spp. which have now been seriously reduced by excessive felling in Australasia, although they once produced fine quality podo and butterbox wood; the latter name explains how this valuable wood was wasted for packaging purposes. The only other southern hemisphere softwood of importance is perhaps parana pine from South Brazil but even this vast resource is now threatened by excessive felling.

In Australasia the Eucalypts are the dominant hardwoods. There are several hundred species with a wide range of properties. They include the giant gums such as jarrah, karri, tallow-wood and many other strong, hard, heavy duty woods. In addition woods such as blackbean and Tasmanian blackwood are decorative as well as being tough, strong and durable. Naturally woods with appearances similar to temperate hardwoods have attracted similar names, so that Tasmanian or Australian silky oak is actually a Eucalypt. In addition there are evergreen "beeches", such as rauli and coigue which were shipped to the British Isles during World War II as beech substitutes.

Altitude

The progressive alteration in forest character that occurs from the tropics to the Polar regions with increase in latitude is imitated with increase in altitude. The snow cap on a mountain represents the Polar region, surrounded by a belt of permanently frozen soil which is unable to support trees. However, the transition at the tree line is rather more abrupt than that caused with change of latitude, perhaps because many mountain ranges are well removed from the Polar region and thus avoid the restricted daylight in winter which is thus a characteristic of latitude but not of altitude. In other respects the altitude and latitude changes are synonymous, passing through the coniferous belt to the temperate, sub-tropical and ultimately tropical forests if the mountain happens to be close to the Equator. Naturally local conditions can modify the situation so that an island has a maritime climate and is perhaps considerably influenced by prevailing winds, entirely different circumstances from the continental climate encountered in the middle of a large land mass. Soil type is also important. For example, in southern England the chalk hills are too porous and too dry to support trees, except where the hills are capped with clay and support beeches. In contrast the alluvial deposits in the valleys support oaks and elms and, where the conditions become swampy, alder and willow dominate.

Virgin forests

In virgin forests, free from the influence of man, a simple process of natural selection decides the species of trees that will grow and strong competition ensures that the growth is very slow. Virgin forests are now extremely rare in the coniferous and temperate hardwood zones, entirely confined to inaccessible mountainous regions in Scandinavia and vast tracks of land far removed from communications in Canada and Russia. Slow grown trees of great age produce wood that is even textured and perhaps more stable and more durable than faster grown wood, yet congested virgin forests are an inefficient method for wood production. The yield per acre of usable wood is comparatively low, partly because mixed species often occur but perhaps because the height of the trees reduces the yield per acre and many of the older trees are, of course, affected by deterioration such as heart rot. Even forest that has been permitted to regenerate naturally after felling is now usually managed by cutting competing herbaceous plants when the trees are young or removing unwanted competing species. In addition it is normal to lop the lower branches from coniferous trees in order to ensure the minimum

Forest management

density of knots in the trunk and progressive thinning is designed to ensure the rapid development of large diameter trees which yield more valuable wood and thus the maximum financial return from the area. Replanting with advantageous species and varieties reared in nurseries is now widely practised over enormous areas, either to improve the yield or quality of wood from natural forests or to generate new forests in wastelands such as sandy deserts, swamps or desolate moorlands which are quite unsuitable for other purposes. Plantation forestry was adopted in France several hundred years ago but in other areas, such as the United States, the British Isles, South Africa and New Zealand, plantation forestry is a comparatively recent development, designed to provide home grown wood at economic costs and thus reduce the need for more expensive imports.

Transportation

Indeed the use of wood originates with fellings in forests surrounding the community, yet wood has been an item of overseas trading for many centuries. Thus straight conifer logs would be transported to temperate or sub-tropical hardwood zones for structural purposes and decorative hardwoods would return for use in valuable furnishings. Exporting centres were generally developed around communications. Thus the coniferous forests of Canada and the Baltic were traditionally felled in the winter months, when the moisture content in the tree is lowest, the logs were then dragged by horses to the river beds and floated down to the exporting ports by the flood from the melting snows. Whilst there is now greater emphasis on the use of surface transport the logs are frequently returned to storage ponds, particularly when the maintenance of a high moisture content is desirable, as in logs intended for peeling for plywood production. Where storage ponds are no longer essential the ability to float becomes unnecessary and, for example, the ringing of teak trees in advance of felling in Thailand has been abandoned as the only purpose was to reduce the moisture content to permit the logs to float. With the exception of some limited use of the railways in North America, ships are invariably used for the long distance transportation of wood so that the exporting ports are almost always situated on suitable rivers or inlets which give access to the seas and oceans of the world. Thus Rangoon and Moulmein, the teak ports of Burma, are situated on the Irrawaddy. In Nigeria logs are rafted down the Niger for shipment from the lower reaches. In the same way the St. Lawrence provides access to the forests of Canada, and the North Baltic, the Gulf of Bothnia, serves the same purpose for the North Swedish and Finnish forests.

Wastage and local mills

It was once traditional to ship wood as logs, but their shape intoduces wastage in shipping space and logs only contain on average 50–60% of useful wood; edging, trimming and rejection of local defects accounts for 35%, whilst sawing removes a further 13%. In order to ensure more efficient utilisation of shipping space sawmills have been widely established around the traditional wood exporting ports, at first in relation to the coniferous forests of Scandinavia and North America but now in Nigeria, Ghana, Brazil, Kenya and Malaya in addition. Plywood, particle board and fibre board mills have been similarly established in the forest zones.

Competing demands for wood

The availability of solid wood is affected by the demand for wood for other purposes, such as the manufacture of plywood, particle board, fibre board, paper and chemicals, as well as the use of wood as a fuel. The use of wood for the manufacture of plywood, particle board and fibreboard is not strictly competitive; the wood is still used for structural purposes and it is obviously sensible for it to be processed so that it is utilised in the most efficient way.

The use of wood as fuel is comparatively insignificant, except perhaps in the under-developed countries where potentially valuable wood may be lost unnecessarily in this way. The amount of wood used for chemical production is similarly insignificant and it is the production of pulp for papermaking that represents the most significant demand for wood. Unfortunately the economic boom of recent years has resulted in vast wastage of paper, or even solid wood, for packaging, and this has created an entirely unnecessary and excessive demand for forest felling. If paper increased in price it might conserve forest resources by discouraging unnecessary wastage and by encouraging recycling of waste paper, a process that is at present uneconomic in most industrial countries.

Market pressures

It is conifers which provide the softwoods, the principal structural woods of the world, that tend to be in excessive demand. Forest plantings have thus been encouraged in many areas but there is also an increasing tendency to utilise alternative hardwoods, some from entirely new areas. For example, there is a vast potential source of virtually unknown hardwoods in the unexplored regions of the Amazon basin and there are active attempts in many parts of the world to develop such alternative sources of wood. Holland perhaps imports the greatest variety of woods of all countries and, in a recent survey of reasonably priced tropical hardwoods, an attempt was made to relate the demand for individual established species with their features. The purpose of the survey was to establish a simplified system for evaluating the market potential of new and unknown species. The survey had some rather unexpected results; the volume demand for individual species could be related to their texture, coarse and medium texture woods being most popular. Price appeared to be directly related to colour, and the only other important factor appeared to be workability. The properties that are often considered to be so important by wood scientists were relatively unimportant in the marketing of wood!

Forest reserves

There have been considerable fears that the demand for wood from certain countries might exceed the forest supplies, or alternatively might increase the price of wood to an unacceptable level. For example, Japan has been suffering from a very serious shortage of structural wood and has been actively expanding its sources of supply. Russia has recently become its principal source of softwoods and Kalimantan (Indonesia) is now its principal source of hardwoods. The Japanese have also attempted to ensure future supplies from Malaya and even Scandinavia, although it appears that these attempts have been largely influenced by panic generated by the fear of declining availability of wood in the future. China is an expanding industrial nation of virtually unknown potential; is China a nett importer or exporter of wood? In fact all these fears regarding the future availability of wood are rather parochial in scope as there are very large forest reserves in areas that are at present relatively inaccessible, such as the hardwoods of the Amazon basin and the softwoods in both Russia and Canada. Even on the basis of existing forest resources the anticipated felling will be well below the natural wastage; European softwood forests have a wastage of 3% per annum but the total fellings are unlikely to exceed 2.5%, whereas in Russia a wastage of about 1.2% is vastly in excess of the 0.5% felling that is anticipated. Another important point is that 67% of the world forest resources are hardwood, yet softwood is at present almost entirely used for structural purposes. The distribution in demand between hardwood and softwoods is clearly outdated and the traditional divisions must be abandoned in the near future, so that light tropical hardwoods will be substituted for softwoods for many purposes. This will naturally result in an increase in demand for tropical hardwoods and the new species that are potentially available from South America require to be more actively investigated.

Efficient utilisation

At the same time it is foolish to increase forest resources and then waste their products. Wood should always be efficiently utilised in order to conserve forest resources, to make more efficient use of the energy and effort that might be wasted on unnecessary working and transporting of wood, and to reduce import costs that are such a serious menace to the economies of many countries. If wood is used in a situation where there is a danger of decay it is essential that durable species should be used or an adequate preservation treatment applied in order to avoid destruction and unnecessary waste. Thinnings, branchwood and tops of small diameter can be combined with trimming and sawing waste to produce particle and fibreboard panel products without the utilisation of any wood that might be suitable for solid use.

Advantages of wood

The purpose of this chapter has been to introduce wood so that its advantages and disadvantages can be fully understood. Its greatest advantage is production as a crop in contrast with other structural materials which are derived from exhaustible mineral resources. The variability between woods of different species may appear to be a disadvantage to the unintelligent user but is, in fact, a distinct advantage as different species have different properties and there is almost always a suitable wood for a particular purpose. The variability that occurs within wood from the same species can be tolerated by increasing dimensions to introduce safety factors, an acceptable procedure in view of the low cost of wood in comparison with competitive materials. Wood is strong with outstanding rigidity in bending and strength in compression. It is resilient, attractive and warm to the touch, easily worked with either hand tools or semi-automatic industrial machinery. It is easily joined with simple fixings such as adhesives, nails or bolts which do not require elaborate tools. Wood has exceptional stability in the longitudinal direction, even when subjected to fluctuating moisture content or exceptional temperatures. Wood is normally exceptionally durable when dry and comparatively inexpensive processes enable it to withstand the most destructive biological agencies, such as insects and fungi. In addition, wood is free from corrosion. Its cellular structure ensures good insulation properties and, whilst it is combustible, it has remarkable fire resistance and maintains its integrity when exposed to serious fires in which metals melt and concrete disintegrates. There is a tendency for engineers and even architects to reject this ancient and natural structural material in favour of the more uniform and more modern manufactured materials but there is no need for them to be suspicious; if they are given as much information about wood as about these newer materials, such as its advantages and the correct way in which it should be utilised, perhaps they will understand it better and will find that it continues to be a reliable structural material when many of the others have ceased to be available and have been forgotten.

2. Converting Trees to Wood in Service

2.1 Conversion

This chapter is concerned with the problem of converting the wood in forest trees into wood in structural service. The term "conversion" has a special significance as it is sometimes applied to the process of converting logs into sawn wood, although this chapter is concerned with the entire process of producing wood from initial forestry to final installation in structures.

Forest management

Forestry represents the first stage in the production of wood. Trees must be selected to provide wood of suitable type for particular uses, either by searching virgin forest for suitable trees, the process still employed in most tropical countries, or culturing suitable trees, a process now widely adopted in many temperate coniferous areas. There are, in fact, two logical systems for cultivation. In the first case a forest is allowed to regenerate naturally after clear felling but the area is tended to produce the best yield of the most valuable wood. Thus the lower branches are removed from trees in order to reduce the number of knots and thinning is practised to encourage more rapid growth of trees to commercial log size. Weed species are removed so that they do not compete for light and for soil moisture with more valuable trees. The second system is the planting of preferred varieties of trees and in this way forests can be extended to land which is otherwise wasted, such as sandy desert or boggy areas, by the selection of suitable species which are able to resist these adverse conditions. In this way forestry can be a direct aid to food production as the presence of a new forest substantially increases the humidity of the atmosphere, creating dew which can enable grass pastures or cereal crops to become established. In more equable conditions the trees are selected to give the highest yield in the economic sense; this does not necessarily mean the highest volume of saleable wood but perhaps occasionally a much smaller volume of very valuable wood, suitable perhaps for decorative veneers.

With a suitable soil and climate a fixed area of forest will produce a fixed volume of wood per year, whatever the density of the tree planting. This is generally the result of restrictions in the lighting of the crown of the tree so that the annual increase in wood in a single tree will depend upon the area that it occupies within the forest. A further characteristic is that generally the height increase in the trees will be maintained, whatever the planting density, and within a certain area the top height will give an indication of the age of the tree. These two factors combined mean that dense planting gives small diameter trees whilst larger trees will be developed with a lighter planting density. In fact it is logical to initially plant trees at a high density and then, when the crowns have filled the spaces and started to compete, a thinning programme can be introduced. In many of the coniferous forests of the British Isles the first thinnings are used as Christmas trees. The second thinnings are far larger and are perhaps sold as pulp wood, pit props or fence posts. The next thinnings yield reasonable sized logs which are used for producing box wood, palettes or chips for particle board manufacture. The thinning programme is designed to produce the highest average yield in the economic sense, and it is largely designed to produce the most valuable saw logs as rapidly as possible, although it is partly influenced by the significant value of the thinnings if there is an appropriate market reasonably close to the forest. In general thinning should only be carried out when there is a danger of stunting and when failure to thin may affect the average yield per year.

Forestry is both a science and an art. The artistry arises in connection with the foresters' ability to rear advantageous species and varieties of trees, and to encourage their growth in perhaps unsuitable conditions. Science arises in the application of the very complicated forest management tables that have been devised to guide the forester with the thinning programme and to forecast the production for individual species so that forests can be managed to produce their optimum yield. Plantations should not be felled until the yield per year per unit area begins to fall; perhaps this is the most accurate definition of maturity. On the other hand, trees must be large enough to provide logs of adequate size for conversion into sawn wood without too much wastage. Generally trees are felled shortly after they achieve the age of maximum annual volume increment as the risk of fungal decay or wind blow damage then progressively increases. In all cases the rotation period is designed to yield the maximum profit, and obviously this can be influenced by the prevailing economic conditions at any particular time. When wood prices are high there will be a tendency to fell early to take advantage of the market whilst a recession will certainly result in delayed felling in the hope that the market will later recover.

Changing quality

In virgin or naturally regenerated mixed forests it is obviously economically desirable to fell large trees as they become mature but this invariably results in substantial damage to surrounding smaller trees. Although the planting of preferred species is much more expensive than natural regeneration it generally avoids growth variations so that the plantation can be clear felled very economically when it ultimately reaches maturity. Unfortunately plantation species and varieties are generally selected for their high volume yield and this is rarely accompanied by good quality. For example, in the northern coniferous forests the thinning and planting programmes have resulted in much more rapidly grown trees with wider annual rings. In addition the economic conditions have suggested that the greatest yield could be obtained by felling these trees as soon as they reach a reasonable log diameter, frequently less than 20 ins. (50 cms.). The result of these policies has been to produce commercial sawn wood with a much higher proportion of sapwood with low durability and high movement, whilst the heartwood has also decreased in durability and stability as a result of its rapid growth. Whilst this has perhaps resulted in softwood from, for example, the Baltic area progressively declining in "quality" in the sense that the texture is becoming more coarse and there is a higher proportion of

Plate 2.1 Mixed virgin forest in British Columbia, the principal source of Douglas fir, western hemlock and western red cedar. *(Council of Forest Industries of British Columbia)*

Plate 2.3 Felling in West Africa using a chain saw, a photograph that gives an excellent indication of the large size of the trees in tropical rain forest. *(UAC Timber)*

Plate 2.2 Birch forest in Finland, utilised principally for plywood manufacture. *(Finnish Plywood Development Association)*

Plate 2.4 Rope-ways on spar trees being used to extract West African logs from the forest and load them on road transport which may take them direct to the mill or to a river for rafting. *(UAC Timber)*

Plate 2.5 A log raft approaching Sapele in Nigeria, where the logs will be converted to sawnwood or plywood. *(UAC Timber)*

sapwood, it has recently been reported that there is an opposite movement in Australia as a result of the establishment of Radiata pine forests. Whereas the traditional slow grown wood from virgin forests has largely been replaced by much faster grown wood in the Baltic region, the Radiata pine forests of Australia are entirely new and a substantial improvement in properties has been observed now that they have become properly established. For example, the compressive strength parallel to the grain is reported to have increased by about 24% since relatively young samples were first tested in 1938 and there have been similar increases in all other strength properties.

Felling and extraction

Petrol driven chain saws are usually used for felling as axes are only economic in a comparatively few tropical countries where labour costs are still very low. Felling is the first stage in the very complex process of converting wood in a forest tree to commercial wood in construction. The cost of felling is comparatively unimportant in many forests where the cost of transportation from the tree in the forest to the ultimate construction site is most significant. Many forests are at present uneconomic as the cost of extraction from the forest is too high, although economic situations change and, as wood becomes a more valuable commodity, the working of these forests may become justified. Horses and bullocks, elephants in Burma and Thailand, wheeled and tracked vehicles, rivers, railways and ships are all involved in the complex transport chain. In some relatively inaccessible forests it is sometimes possible to leave "spar" trees or to erect steel spars to support rope-ways to haul the logs to the nearest road or river. In fact rivers were once the most popular form of transport between the forest and the port or saw mill. In the northern coniferous forests it was traditional to fell during the winter months, accumulating the logs on the frozen rivers so that the thawing snow in the spring would convey them naturally towards the mills and ports. Unfortunately this system limits the catchment areas which are effectively confined to the slopes adjacent to reasonably large rivers but, even when road transport is used for the initial extraction from the forest, logs are sometimes rafted considerable distances down rivers. If road transport is used for the entire distance it is quite common in almost all parts of the world to discharge the logs into ponds where they can be stored and conveniently sorted for processing. In tropical countries floating protects the logs from splitting and also from insect attack, although it is fairly common to find the occasional log which, because of its eccentric shape, has always floated with the same side upwards so that limited insect borer attack has occurred.

Transport economics

Excessive volume or weight must be avoided throughout the transportation operations. Round logs are very wasteful in terms of space, both because of the gaps that naturally occur when logs are stacked but also since logs contain only 50 to 60% of useful wood; about 13% is lost by sawing and about 35% by edging, trimming and rejection due to local defects. Wet green wood is very heavy as it commonly contains 60 to 200% moisture. In addition it is particularly susceptible to sapstain discolouration. There is thus an increasing tendency to transport logs for the shortest possible distance between the forest and the saw mill. Whereas wood was once almost universally shipped in log form to mills in the user country, it is now much more usual for the saw mills to be sited close to the productive forests. In

Integrated conversion mills

many cases large centralized mills have been established with either direct loading of sawnwood onto ships or a convenient transport route to a port. The mills are often highly sophisticated, accepting logs of all sizes and species but sorting them so that the most suitable are used for sawnwood whilst others are diverted for plywood or particle board production. In other integrated installations a sawnwood mill may be combined with perhaps a pulp and paper mill, or even chemical installations producing cellulose. Obviously these integrated mills are designed to utilise the forest crop in the most efficient manner possible but they also represent considerable capital investment and must be utilised continuously if they are to be profitable. It was traditional in northern coniferous forests to fell trees and extract logs in the winter when they can be hauled from the forest on the frozen ground. This also had the advantage that the moisture content was lowest in the winter and there was no danger of insect or fungal damage. A comparatively short period of winter felling, combined with limited scope for storage, ensures that most installations have now been forced to adopt felling throughout the year and this is naturally associated with increased moisture content and an increased danger of insect and fungal attack.

Converting logs to sawnwood

Logs arriving at mills in northern coniferous forests are stacked in a yard or floated on a storage pond. They are then carried to the upper level of the mill on a "jack ladder" consisting of a moving belt with teeth or spikes to grip the log. This upper level of the mill contains the conversion saws, "converting" the log to sawn wood. In Northern Europe it is normal to use reciprocating frame saws with multiple blades that can be adjusted to give the required width of boards. Usually the first cuts are on either side of a log to give a centre block and sideboards. These sideboards are often diverted by rollers to the edging saws, although they are largely waste and in some mills they are diverted to the lower level and perhaps into another section of the mill where they are used for the manufacture of particle board or paper pulp. The centre block is turned on one of the cut sides and passed through a second frame saw to produce "deals" or "battens"; the various sizes of converted wood will be described later in this chapter. Some circular or rotary saws are used in Europe, usually in smaller mills or for resawing to smaller sizes. The trees on the west coast of the United States and Canada produce very large logs which are normally converted using band saws, although frame and circular saws are frequently used for smaller logs. Conversion sawing is followed by trimming to lengths and grading, and then the wood is either dried in a kiln or treated with anti-stain chemicals before being moved to the storage area. In many mills the sideboards, edge trimmings and sawdust are used for firing furnaces to provide heating or even steam power for the mill. The tops of the trees that are too small for use as logs are, of course, removed in the forest but even these are often transported to suitable mills for conversion to chips or pulp; the present cost of wood and the pressure on forest resources means that nothing must be wasted.

The manner in which the log is sawn is usually considered to be relatively unimportant. The simplest technique is to make a large number of parallel cuts, a method known as through-and-through, flat sawn, back sawn or plane sawn. The outer boards are cut largely in the tangential plane whilst the middle board is in the radial plane, the angle between the annual rings and the surface of the board progressively varying through the intermediate boards. In the tangential boards there is a tendency to cup through a change in moisture content as one face of the board is towards the bark and the other towards the heart of the tree. Cupping is not too severe when the board is entirely heartwood when prepared from a large log, but cupping can be very severe if the board is cut through the interface between the heartwood and sapwood. There is also a tendency for splits to occur along the spring wood of an annual ring, causing the inner rings to "shell out". In flat sawn boards

Figure 2.1 Flat sawn logs.

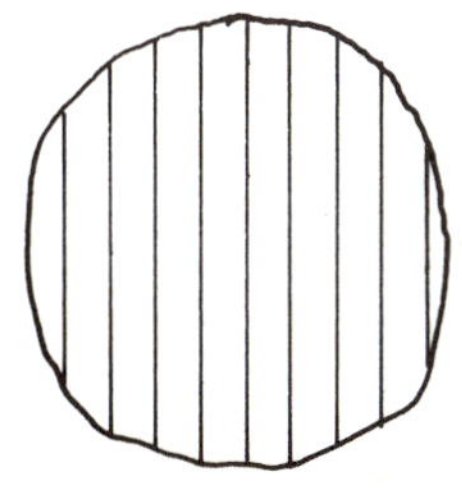

the centre board is weaker where it incorporates the pith and specifications sometimes specify "free from heart" to avoid this weakness. In fact, the pith can also be the origin of some fungal decay defects and when these are obvious the sawyer may "box the heart" of a log; sawing so that the defective area around the pith can be rejected. There are also various techniques for preparing boards that are predominately in the radial plane, known as quarter sawn, rift sawn or, in North America, vertical grain (v.g.). These boards show the annual rings as regular parallel lines on their faces and are therefore known as edge grain boards. Quarter sawn boards are particularly suitable for use as flooring as they do not suffer the cupping that is a characteristic of tangential or outer flat sawn boards. The edge grain on the surface also gives good wear resistance. As radial movement is usually far less than tangential movement these boards are stable in width whilst the movement in thickness is insignificant, except perhaps when sapwood is incorporated. Quarter sawn boards are very expensive to prepare because of the excessive amount of handling on the saw bench and the large amount of wastage, particularly through waney edge losses arising because of the taper of the log.

Figure 2.2 Quarter sawn logs.

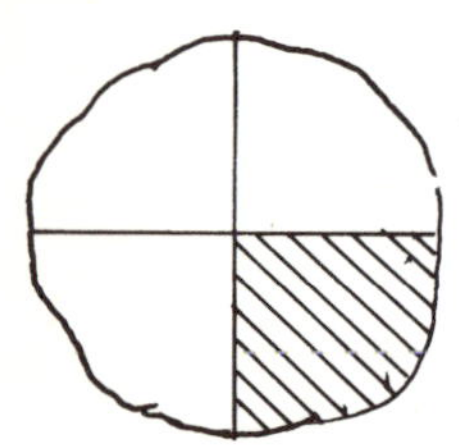

Tropical hardwoods are generally converted using band-saws situated on the same level as the storage yard; the logs are often too heavy to be readily lifted to an upper floor in a mill. Logs are converted either for maximum yield of sound wood or alternatively to develop the most valuable decorative figure. For example, interlocking grain will yield stripe or ribbon figure in quarter sawn mahogany, and silver figure will only occur if oak is also quarter sawn. Both tropical and temperate hardwoods are often flat sawn through and through and then the boards are subsequently restacked in their original order, properly "stickered" or with the boards separated by small sticks to permit drying, an air seasoning arrangement known as a "boule". This system is only realistic where the direction of cut is relatively unimportant. An alternative method of conversion is to make three through-and-through cuts to provide two flat sawn boards from the centre of the log. These will naturally include any heart defects which can then be removed when the boards are resawn. The remaining wood consists essentially of two half logs, often known as wainscot billets. These billets are then turned onto their flat face and resawn to give a number of boards. These naturally vary in width between the centre and edge of the billet but their faces display a relatively large proportion of edge grain or predominantly quarter sawn wood and the wider boards in particular develop the characteristic figure arising from this direction of cut. This method of conversion is known as wainscot sawing because it was developed for the conversion of oak for wainscoting or panelling of rooms; the varying widths of the boards and the silver figure in the quarter sawn wider boards is particularly attractive for panelling, whilst the wider boards are free from cupping and possess the lowest movement as they are quarter sawn.

Figure 2.3 Hardwood boards in boule.

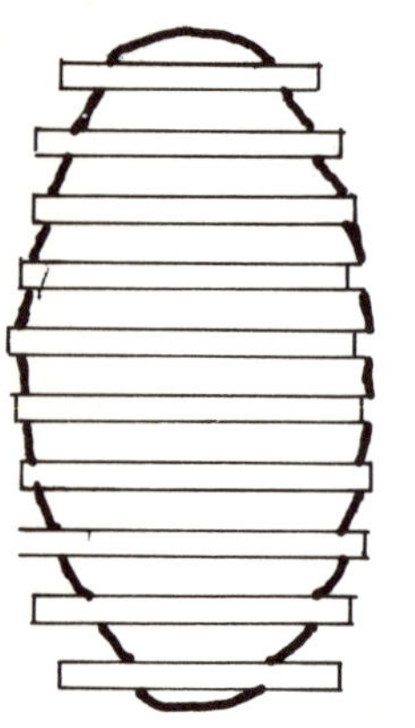

Figure 2.4 Wainscot sawn logs.

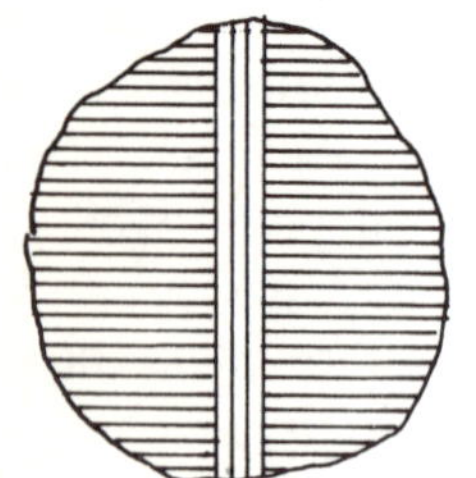

It is essential that saw milling operations should achieve the highest possible utilisation of wood if forest resources are to be properly conserved. Pieces that are too small for use as commercial sawn wood must be utilised, together with small diameter logs, for the manufacture of pallets, boxes, chips and pulp. Waste wood and saw dust is now rarely used for firing furnaces for steam raising, although it is still widely used for space heating, except in mills where there is a ready use for saw dust and smaller offcuts in pulping for the manufacture of paper or fibre boards. Unfortunately wood is frequently resawn to produce pieces of smaller cross-section, perhaps at the shipping terminal or a mill in an importing country in order to satisfy a demand for smaller sizes, but this is invariably a wasteful process; it is now normal practice to fell softwood trees in particular when they are of a comparatively small diameter so that it is normally difficult to obtain wood in larger sizes. Despite this disadvantage resawing is widely practised, although it involves additional costs in both resawing and perhaps retreating to prevent subsequent stain development on the freshly sawn faces.

Plate 2.6 **The headrig, a band-saw and table used for cutting the log into baulks.** *(UAC Timber)*

Plate 2.7 **A frame saw, used for converting baulks into boards.** *(UAC Timber)*

Milling, planing and moulding

Milling is the next stage in the preparation of wood for service. Whilst a sawn surface may be perfectly satisfactory for structural wood, a smooth planed surface is required for all applications in which the wood is to receive a paint or varnish finish, or where it is likely to come into direct touch contact with people. Generally this means the entire non-structural wood in a building, including the skirtings, floorboards and the joinery (millwork) consisting of doors, door frames, door stops, architraves, skirtings and panelling. External joinery (millwork) consists of door and window frames and surrounds, facias, barge-boards and cladding. In its simplest form milling consists of the preparation of a "planed square edged" board (p.s.e.) but boards can be finished in a variety of more elaborate forms such as "planed tongued and grooved" board (p.t.g.) or "planed tongued and grooved with V joint" board (p.t.g.v.) which is also known as match boarding. Special sections are prepared for cladding, including plain weather board, rebated weather board and shiplap, although p.t.g.v. or match boarding is often used as a vertical cladding. An infinite variety of mouldings are prepared for architraves, skirtings, sills, drips and various window and door frame sections. The simplest milling machinery consists of a planer which will prepare each face separately by passing the wood beneath a rotating shaft on which a number of plane blades are fixed. A thicknesser has blades both above and below the wood whilst various combination machines may have many shafts or spindles to accommodate either plain or shaped cutters. A four-cutter can prepare all four faces of a piece of wood in a single operation whilst additional spindles can be provided, usually up to seven on an individual machine, for special work, such as the cutting of special rebates, grooves or champfers. The production of a combination machine can often be doubled by preparing two mouldings back to back so that they are passed through a splitting saw as the final cutting operation, in order to separate them. These more sophisticated combination machines are only economic for long production runs as they are expensive to adjust for an individual operation and the maintenance of the cutting spindles can involve considerable work in sharpening and balancing, particularly when abrasive tropical woods are being processed. For short production runs it is usually more economic to employ a spindle moulder, consisting of a single vertical spindle in which a single cutter can be mounted, so that a single operation on one face of a piece of wood can be completed with each pass of the cutter.

Figure 2.5 Typical board sections.

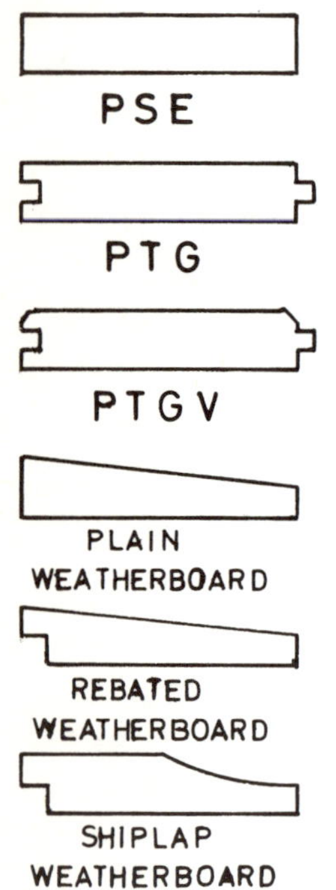

Wood merchanting

All these various operations are handled by a variety of firms, commencing with the shippers who are actually responsible for exporting wood. They may own their own forests and saw mills or alternatively have special arrangements with groups of mills in the vicinity of their exporting ports where they may own their own extensive terminal facilities. The shippers appoint agents to represent them in overseas markets. The function of an agent is to sell the wood offered by the shipper, and the agent is usually paid by a commission on these sales. In addition many agents also advance capital to the shipper to enable them to accumulate stock. The agent arranges the sale with an importer who generally contracts to buy wood many months in advance of delivery. The importer owns or rents extensive facilities so that large shipments can be handled and distributed in turn to merchants. Brokers are also involved, introducing buyers and sellers at various stages in these operations.

Seasonal fluctuations in supplies

Exporting operations are often seasonal and the ports in the North Baltic, White Sea and Kara Sea are, for example, ice bound throughout the winter months so that wood exports cannot commence until "first open water" (f.o.w.). Ice breakers are now frequently used to extend the shipping season but it will be appreciated that it is still necessary for importers to maintain large stocks in order to satisfy a continuous demand from sources with seasonal fluctuations. In addition it is necessary to purchase wood well in advance so that the supplies can be prepared and accumulated at the exporting port for shipment when conditions permit. If the economic conditions indicate a continuing demand for wood and thus a sellers market, the agents and importers may be willing to commit themselves to contracts perhaps nine months before shipment, but in times of recession with high interest rates they will be rather more reluctant to purchase in advance. This introduces a considerable problem for the shippers who must then decide for themselves whether to accumulate stocks in anticipation of future orders or whether to reduce their rate of felling and conversion, perhaps missing lucrative late season orders.

Transport

At one time all sawnwood was shipped loose in random lengths but softwood and large quantities of tropical hardwoods are now shipped as packages containing pieces of fixed lengths. Whilst the dominant system of transportation is by ships where large distances are involved, there are many instances of direct railway links between the productive forests and the user areas, particularly in North America. In Europe there is a tendency to develop direct road links between the Scandinavian countries and the principal European industrial countries. Whilst road transport is expensive over long distances, particularly where ferry links are involved, the system gains by delivering direct from a forest mill to a merchant's yard, avoiding the many handling and storage operations involved with the conventional shipping system. Packaged wood is much easier to handle than loose pieces, provided suitable machinery is available, but necessarily involves standard length pieces. There is obviously a need to avoid waste in cutting to lengths and the logical alternative is to continuously join end-to-end the entire production of a mill, subsequently cutting to the lengths required for packaging; the use of finger joints for this purpose is discussed later in this chapter.

Grading

After wood has been converted and sawn to lengths it is sorted into various grades representing its suitability for different purposes. In the past sorting was entirely visual and it was often only related to the strength properties of the individual piece of wood in a rather arbitrary manner. For example, Baltic redwood from Sweden was sorted into five grades, the following table indicating the allowable defects for deals 3 in. × 9 in. × 17ft:

"Firsts". No blue stain, shakes to $\frac{1}{4}$ in. deep, 3 or 4 knots under $\frac{7}{8}$ in. diameter.

"Seconds". Slight blue stain, shakes to $\frac{1}{2}$ in. deep, 3 or 4 knots under $1\frac{1}{4}$ in. diameter.

"Thirds". Blue stain in one edge, shakes to $1\frac{1}{4}$ in. deep, 5 or 6 knots under $2\frac{3}{4}$ in. diameter.

"Fourths". Blue stain up to 2/3rds overall, shakes to 50% depth, 6 or 7 knots under $3\frac{1}{4}$ in. diameter.

"Fifths". Blue stain all over, any shakes, any knots.

Stress grading

Naturally the limitations on shakes and knots meant that firsts were stronger than fifths but the emphasis on blue stain was largely unnecessary for structural wood. For this reason there has been a tendency in recent years to introduce stress grading for the sorting of wood for structural purposes. The grades of individual pieces must be marked on them so that it is possible to identify pieces of wood that are considered to be suitable for particular structural purposes. Sapstain, or blue stain, is largely ignored; this is considered to be a later selection problem where wood is required free from stain for some decorative purpose. Stress grading is thus concerned only with the factors that affect strength. Knots are the most significant single factor but, whereas the older systems of grading were concerned largely with the numbers and sizes of knots in an individual piece, stress grading takes account of the type of knot and its position. If the knot arises as a result of cutting through a living branch, it is firmly incorporated within the trunk wood of the tree and the main defect is disorientation of the grain; this defect is known as a live knot. If a branch has died and becomes surrounded by subsequent trunk tissue it produces a dead knot surrounded by a ring of bark, so that it is likely to fall out as the wood dries. For this reason forestry involves the trimming of the lower dead and drying branches close to the bark of the trunk in order to avoid the development of dead knots. In some cases fungal decay will have progressed down the dead side branch to produce rotten knots. The form of the knot is also important. If cross-cut across the branch a round knot is produced, whereas a longitudinal cut through the branch gives a splay knot and an oblique cut will produce an oval knot. Other significant factors in stress grading are the slope of the grain, the rate of growth and size deviation in conversion, as well as the presence of splits or fissures, wane, resin pockets and distortion. Probably the best way to appreciate stress grading principles is to consider a particular system in detail.

The Building Regulations for England and Wales were amended with effect from the 31st January, 1975 to require each piece of wood used in a strength application in buildings, even in houses constructed traditionally, to be stress graded and marked according to the British Standard Specification B.S. 4978: 1973 "Timber grades for structural use". Wood stress graded to other rules and grades would be acceptable but only if special approval had been given for its use; for example, the Canadian NLGA rules have also been accepted. Obviously the grading and marking of individual pieces of wood is alone insufficient and the building regulations also require that the pieces should be suitable for their use as defined in the British Standard Code of Practice, CP 112: 1971 "The structural use of timber", including amendment 1265. It will therefore be appreciated that the reliable stress grading of wood and the marking of all pieces is now essential, and Baltic redwood exported from Sweden to the British market is now graded and marked in accordance with B.S. 4978.

Visual grading

The British Standard Specification B.S. 4978: 1973 "Timber grades for structural use" permits approved species of softwood to be graded visually or by machines. There are two grades for solid wood, known as General Structural (G.S.) and Special Structural (S.S.), the equivalent machine grades being M.G.S. and M.S.S. respectively. The first feature in grading is to ensure that the pieces of wood possess their proper cross-section dimensions as any loss of dimension, perhaps through resawing, will result in a loss of strength; the permitted tolerances are defined in a separate British Standard Specification, B.S. 4471 "Dimensions for softwood". The presence of wane, the curved edge of the log, along the arris of the piece of wood also represents a loss of cross-section and thus a loss of strength; the area lost by wane from an individual face of the piece must be less than one quarter for S.S. grade and one third for G.S. grade of the total area of the face, and in G.S. grade within 300 mm. of the end of the piece up to half the width can be lost in wane, provided the continuous length of wane is less than 300 mm.

The most important factor in grading to B.S. 4978 is the Knot Area Ratio, defined as the ratio of the sum of the knot area projected on a cross-section relative to the total area of the section. The point of this requirement is that it is considered that the integrity of the cross-section throughout the length of the piece is perhaps the most important factor in terms of strength. When a piece is used as a beam the upper wood is in compression and the lower wood in tension. The margins of a piece of wood are defined as the outer quarters of the width of the piece and when knots occur in these margins they clearly have a greater significance on the strength of a beam than when they occur in the relatively unstressed centre of the beam. A margin condition is said to exist when more than half the area of either margin is occupied by projected knots. G.S. grade is then defined as wood with a Knot Area Ratio of less than one half, or, if margin condition exists, less than one third; margin condition is not applied to square section pieces for which the Knot Area Ratio should be less than one third. S.S. grade has a Knot Area Ratio of less than one third or, in the case of margin conditions or square sections, less than one fifth. In estimating the Knot Area Ratio for grading purposes all knots of less than 5 mm. diameter are ignored and there is no distinction between empty knot holes, dead knots, or live knots. The longitudinal separation of the knots is however, important and for both G.S. and S.S. grades the piece must be rejected where two or more knots or groups of knots with a Knot Area Ratio of more than 90% of the total permitted are separated in length by less than half the width of the piece.

Figure 2.6 **Knot Area Ratio (projected knots are shaded).**

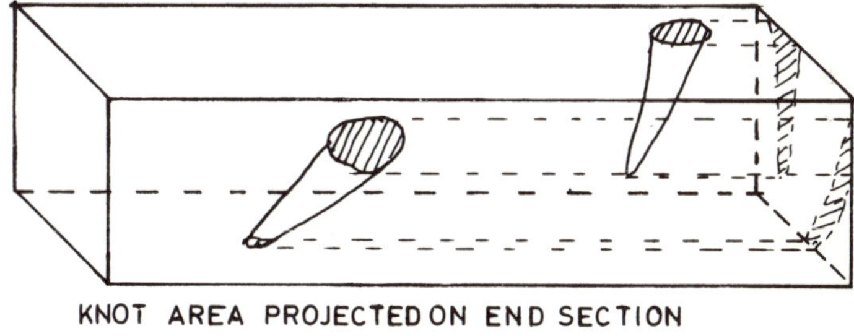

Figure 2.7 Margin condition (projected knots are shaded).

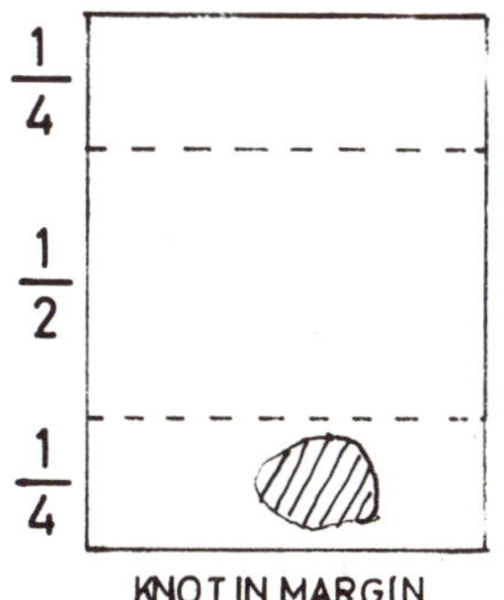

This grading system is obviously extremely complex but it can be operated reasonably reliably if the grader adopts a logical sequence of decisions:

Knot Area Ratio (BS. 4978)

less than $\frac{1}{5}$ .... SS grade
less than $\frac{1}{3}$ ....no margin condition................ SS grade
...............margin condition.................... GS grade
less than $\frac{1}{2}$ ....no margin condition................ GS grade
...............margin condition.................... reject
more than $\frac{1}{2}$ .... reject.

Splits, fissures and resin pockets must also be taken into account. Unlimited fissures are permitted if they are less than one half the thickness of the piece. If they are greater than half the thickness but less than the whole thickness their length should not exceed 900 mm. for G.S. grade or 600 mm.

for S.S. grade or one quarter of the length of the piece, whichever is the lesser. If the size of the defect is equal to the thickness the length shall not exceed 600 mm. for G.S. grade or, if the defect occurs at the end of the piece, its length should not exceed 1.5 times the width of the piece; in S.S. grade fissures equal to the thickness are only permitted within one width length of the end of the piece.

Figure 2.8 Slope of grain (1 in X).

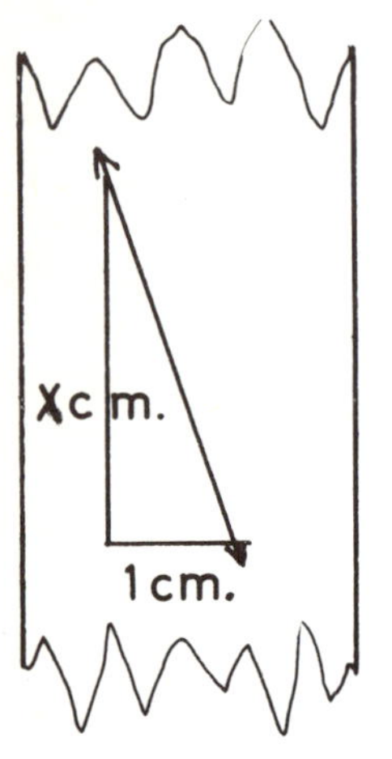

The growth rate should not be excessive and there should not be less than four rings per 25 mm. In addition the slope of the grain should be less than 1 in 6 for G.S. and 1 in 10 for S.S. grades; the angle of grain is determined by pulling a needle on the end of a shaped rod across the surface of the piece of wood. A method which is perhaps simpler, although not mentioned in B.S. 4978, is to place a spot of Indian ink on the surface of the wood, as it will spread along the grain, indicating its slope.

Finally the wood should not be unduly distorted. Within 3 metres bow should not exceed one half the thickness, spring should not exceed 15 mm. and twist should not exceed 1 mm. per 25 mm. of width. In addition the depth of cupping should not exceed one twenty-fifth of the width. Some small bore holes, such as pin holes, are permitted provided there is no evidence of current activity, but larger holes, such as those produced by wood wasps, must result in rejection of a piece. Fungal decay damage and other abnormalities will also lead to automatic rejection. Sapstain is not structurally significant and is not considered in stress grading.

B.S. 4978 also includes a simplified grading system for wood to be used in laminated structures. The main relaxation from the general grading system is to ignore the margin condition which is much less significant in laminated structures; indeed one of the great advantages of lamination is to utilise relatively low grade wood in the production of high grade structures. The laminating grades are L.A., L.B. and L.C. with Knot Area Ratios less than one tenth, one quarter and one half, and with a slope of grain of less than 1 in 18, 1 in 14 and 1 in 8 respectively.

Machine grading

In machine stress grading the wood is passed between rollers which deflect it and automatically measure its elasticity in bending. Clearly visual stress grading according to these very complex British Standard rules is only economic where the graders are very experienced, but grading machines have the advantage that they only require conscientious operation rather than extensive grading experience.

It has already been explained that the Building Regulations for England and Wales permit the use of other approved grading systems, including the Canadian NLGA rules. In fact, most stress grading schemes adopt the same principles that have been described for the British Standard scheme but they are sorted into different divisions. The Canadian NLGA rules sort into Structural Grades, known as Select, number 1, number 2 and number 3 structural. In addition there are Light Framing Grades, known as Construction and Standard, and Utility. Generally this type of grading system is used throughout North America and there is, for example, uniformity on the West Coast between the West Coast Lumberman's Association in the United States of America and the British Columbia Lumber Manufacturers' Association in Canada, both using rules published by the Pacific Lumber Inspection Bureau. Obviously these modern grading rules are only concerned with the use of wood in construction and they largely ignore any feature that does not have a significant effect upon strength. In some applications, such as in the production of joinery (millwork) sapstain may be unacceptable or a compact fine textured wood desirable in order to achieve a satisfactory finish. Joinery (millwork) is not a use for which high strength is esssential but obviously appearance and woodworking properties are particularly important. The very slow grown woods from the far North are far more suitable for these purposes and it is still common to find specifications requiring all joinery redwood to be from, for example, the Kara Sea or the White Sea, in an effort to ensure that the wood is fine textured and comparatively durable.

Plate 2.8

A testing machine for bending strength. This particular type of machine is designed especially for testing finger joints. In machine stress grading a similar principle is involved, each piece being loaded so that the deflection indicates the stress grade. *(Cook Bolinders Ltd.)*

Cutting system for hardwoods

Hardwoods are often visually graded using a "cutting" system; in principle this system is based upon the amount of a board that can be cut out and utilised. For example, under the Malayan grading rules a board has a number of "cutting units" equal to its length in feet times its width in inches, so that a 20 ft. by 18 in. plank consists of 360 cutting units. The board is then marked with the rejected areas, such as knots and end shakes, and the grader then decides how it can be usefully cut into smaller planks. The superficial area of these planks is assessed in cutting units and the total compared with the original total for the plank to give a grading ratio. The normal grades and their ratios are Prime (10/12ths), Select (9/12ths), Standard (8/12ths) and Serviceable (6/12ths). In Prime, Select and Standard there must not be more than five smaller boards involved in calculating the cutting and the number must usually be stated. There are also a minimum number of sizes for the cuttings. Wood graded in this way is usually marketed as Select and Better, Standard and Better, and Serviceable and Better, the latter also known as Sound and Better, or Merchantable.

Figure 2.9 **Hardwood Grading Ratio**

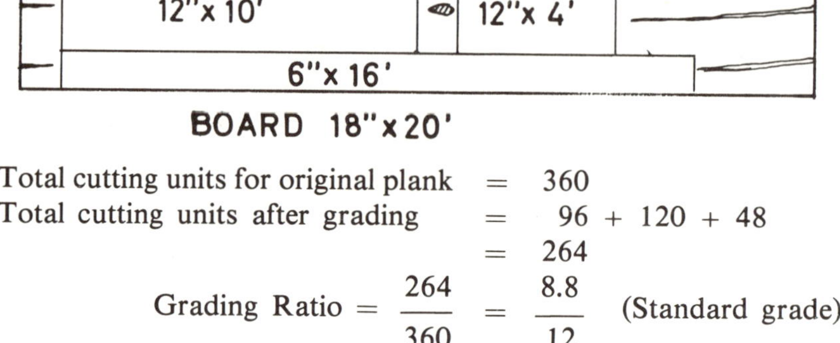

Total cutting units for original plank = 360
Total cutting units after grading = 96 + 120 + 48
= 264

$$\text{Grading Ratio} = \frac{264}{360} = \frac{8.8}{12} \quad \text{(Standard grade)}$$

Stain in softwoods

Although sapstain or blue stain has little structural significance its presence generally limits the usefulness of wood to non-decorative purposes and thus effectively reduces its quality. Thus, whilst staining may not be included within the grading rules, agents and importers may specify that the wood must be free from stain. Most softwood receives an anti-stain treatment immediately after grading, usually by immersion in a suitable fungicide solution such as 1 to 3% sodium pentachlorophenate. Alternatively the wood can be rapidly kiln dried to prevent stain. There are many people who dislike treatments which involve extensive use of toxic chemicals and it was therefore suggested several years ago that all Swedish exports should be kiln dried to avoid stain damage. In fact, it is frequently found that stain can develop during the early period in the kiln whilst the moisture content of the wood is still high and whilst the temperature is too low to kill the staining fungi. In addition it has also been found that kiln dried wood will stain if it becomes rewetted, and finally the energy crisis of recent years has caused an enormous increase in kiln operating costs. Actual practice has therefore shown that wood may need to be anti-stain treated both before and after kilning, and there is now a general tendency for softwood for structural purposes to receive only an anti-stain treatment without any attempt to dry other than normal storage prior to shipment. However, this system is only realistic if a clear distinction can be made between sawnwood intended for general structural use and specially selected wood for joinery (millwork). This special wood is often anti-stain treated both before and after kilning and the final dry product is wrapped in plastic sheeting in order to prevent it rewetting during transportation to the joinery mill. Unfortunately the wood still contains an appreciable amount of water, generally about 12%, and this is sufficient to result in condensation under the plastic sheet during wide temperature fluctuations. Staining of the outer pieces in the wrapped package is therefore comparatively common unless the wood has been anti-stain treated.

As most softwood mills now produce packaged wood there is an increasing tendency for this to be anti-stain treated in the bundle rather than as individual pieces. In fact the tight strapping prevents the penetration of the treatment which is largely superficial and the staining of pieces deep in the package is unfortunately comparatively common. Whilst it must be accepted that the immersion of packages is very convenient for the mills there are many systems that will enable them to treat individual pieces without increasing their application costs. For example, the production can be fed onto a chain conveyor which will carry the pieces of wood into a treatment bath before packaging. Apparatus of this type is fairly expensive and it is probably more realistic to install comparatively inexpensive spray coating machines on all production lines as they leave the graders. There have been attempts to introduce mist coaters which apply a concentrated fungicide to the surface of the sawn wood but staining problems often arise in wood treated in this way, apparently because of the limited penetration of the treatment. Normal spray coaters or deluging machines would be more efficient but they have rarely been considered for this use, although they are widely used for the application of other types of wood preservative.

Stain and pinhole borer in hardwoods

In the tropics there is a considerable danger of both stain and insect attack immediately a tree is felled. The danger is reduced if logs are debarked promptly and then sprayed with a combined insecticide and fungicide to control pinhole borers and stain fungi. Alternatively logs which are immediately floated in a river are relatively free from attack, although sometimes logs float with the same limited area of bark above the surface, enabling some attack to occur.

Moisture content and seasoning

Shipping specifications often require wood to be "properly seasoned for shipment to the country of destination". This generally means that the moisture content of the wood must be sufficiently low to ensure that no deterioration occurs within the hold of the ship. In theory this means that the wood must be dried in some way to a moisture content below about 22%, compared with an average moisture content of 60 to 200% when freshly felled. Seasoning is the term that is generally applied to the various processes that are adopted for reducing the moisture content. Various shrinkage defects can develop if wood is dried too rapidly, so that much of the expertise in seasoning is devoted to achieving the maximum drying rate whilst still avoiding the development of defects. Shrinkage and swelling will also occur as the moisture content varies in service so that seasoning should attempt to dry the wood to the average moisture content that it will achieve in final use so that movement defects can be kept to a minimum. Wood which is protected from the rain but in the open in a temperate environment will have an equilibrium moisture content of about 18%. In a normal centrally heated atmosphere this will be reduced to 12%, and only 8% for blocks and boards installed over floor heating systems. In climates with very cold dry winters the moisture content may fall to only 4%, yet the summer moisture content may be as high as 12% introducing a considerable danger of movement defects unless stable low movement woods are utilised.

Moisture content measurement

The first stage in controlling the drying of wood is to adopt a system for measuring its moisture content. In the normal oven drying method a piece at least 9 in. (22.5 cm.) is cut from the end of the piece of wood to be tested, followed by a $\frac{1}{2}$ in. (1 cm.) thick slice which is the sample for actual test. This sample slice, which must be free from knots or any other obvious defects, must be weighed immediately and then transferred to a ventilated oven with a temperature of 101 to 105 °C (214 to 221 °F.). The sample is removed at intervals for weighing until there is no further loss of weight, usually after 6 to

24 hours. The moisture content is then calculated according to the following formula:

$$\text{Moisture content (\%)} = \frac{\text{(initial wet weight)} - \text{(final dry weight)}}{\text{final dry weight}} \times 100.$$

Obviously the oven will contain a number of different samples at any time but it is important that a new wet sample should not be introduced to the oven just before the removal of another sample for the determination of its final dry weight. This method for determining moisture content is very accurate and reliable but the time involved makes it entirely unsuitable for following the moisture content of, for example, a batch of wood in a kiln during a drying cycle. This difficulty is avoided by selecting one or more pieces of wood from the kiln charge for the preparation of samples for monitoring the kiln operation. The first 9 in. (22.5 cm.) or more is rejected from the end of the sample and then a slice cut as previously described for the determination of the moisture content by the oven drying method. This is followed by cutting a length of wood about 6 ft. (2 metres) in length, followed by a further slice for a second moisture content measurement. The 6 ft. (2 metre) length of wood is end sealed and then weighed; this kiln sample is then considered to be at a moisture content similar to that determined for the two slices taken from either end. The kiln sample is then returned to a suitable position within the stack in the kiln and is removed at intervals for weighing so that the drying of the wood in the stack can be readily followed.

Figure 2.10 **Kiln sample for moisture content determination.**

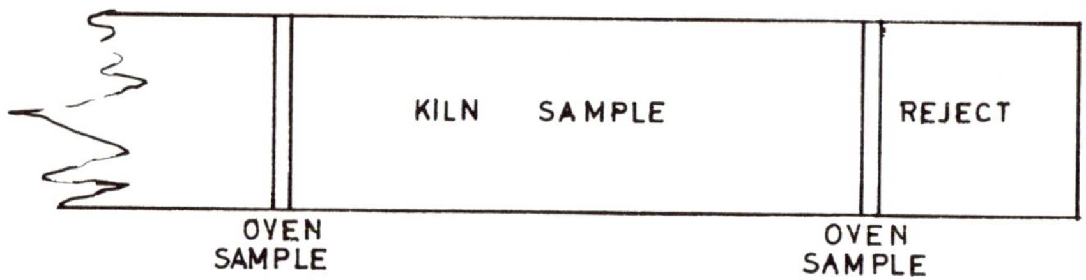

If it is necessary to determine the moisture content more rapidly the oven dry method can still be used by preparing four $\frac{1}{8}$ in. (3 mm.) slices in place of each of the $\frac{1}{2}$ in. (1 cm.) slices previously mentioned. In addition it is sometimes necessary to check the distribution of moisture within the cross-section of a piece of wood, and this is achieved by cutting out a piece perhaps 2 in. (5 cm.) long and then slicing this into separate pieces representing the inner and outer zones for the moisture content determination.

A suitable ventilated oven can be constructed using light bulbs as the heating system. A thermometer is also required for the oven and the temperature can normally be controlled by varying the ventilation. A students' chemical balance will give sufficient accuracy for the weighing although semi-automatic and automatic balances are obviously an advantage where a large number of weighings are likely to be involved. A further large balance will also be required for weighing the 6 ft. (2 metre) kiln samples where this method is used for monitoring the drying of a charge in a kiln.

The main disadvantages of the oven drying method are the fact that the sample is destroyed and it takes a considerable time to obtain the moisture content result. The time delay can be avoided by using a carbide moisture content tester. In this system a sample of perhaps 3 or 6 g. is taken by drilling into the piece of wood to be tested. A twist drill or a small diameter woodworking bit will give a sample that is sufficiently finely divided for this test. This sample is then weighed and placed in a pressure vessel where it is mixed with calcium carbide. The moisture content of the sample is determined by measuring the pressure of the acetylene gas generated by the reaction between the calcium carbide and the water in the sample. Certain precautions are necessary to ensure that the carbide test gives a reliable result. A high speed drill should not be employed as it generates heat which reduces the moisture content of the sample, but the drill must produce a finely divided sample if it is to achieve proper mixing with the carbide, although pulverizing balls are often provided in the pressure vessel which will crush the sample with the carbide. The carbide and sample must not be mixed before the pressure vessel is completely sealed. In some instruments this is achieved by introducing the carbide in a glass phial which is broken by the pulverizing balls when the instrument is shaken after sealing. In other instruments the carbide is placed in the main body of the instrument whilst the sample is placed in a cap, and the two are only mixed when the instrument is inverted after sealing. It is important to allow sufficient time for the reaction to be completed and for the pressure to stabilise before taking the measurement. One final feature of the carbide test is that most instruments normally measure moisture content as a percentage of the wet weight, and a conversion is necessary to express this in the more normal manner as a percentage of dry weight.

Finally, electronic moisture meters provide the most convenient method for the instantaneous and non-destructive measurement of moisture content. In most instruments needle probes are forced into the wood and the electrical conductivity between them gives an assessment of the moisture content. It is important that the probes should be orientated along the grain and the instruments must be calibrated for different species of wood. One disadvantage of this method is the insensitivity of these instruments at moisture contents above perhaps 30% but they are certainly the most convenient means for following the progress of drying in a kiln during the critical stages below the fibre saturation point. Their only disadvantage lies in the fact that they only measure the moisture content to a short distance beyond the depth of penetration of the needles, and perhaps the most reliable method for following the operation of a kiln is to use a combination of an

Plate 2.9 **Protimeter Timbermaster electronic moisture meter with a hammer electrode which enables the probes to be forced into hard dense woods.** *(Protimeter Ltd)*

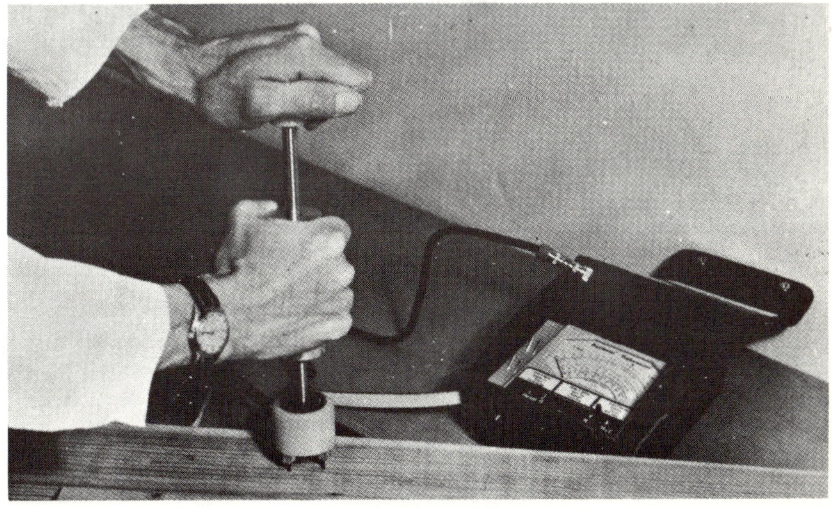

electronic moisture meter of this type together with a large kiln sample which can be weighed at infrequent intervals in order to check the results obtained with the electronic moisture meter.

Air seasoning

The traditional method for seasoning or drying wood is to stack it in the open air, although it will be appreciated that drying will only be achieved if the stacks are protected from the rain, either by providing a roof to each individual stack or by the use of large open sided seasoning sheds. Ground contact must be avoided, and grass and weeds controlled so that they do not prevent the drying of the lower wood in the stack. One serious problem with air seasoning is the excessive drying rate during hot weather. This can be a very serious problem as drying occurs most rapidly from the porous end-grain which thus shrinks in advance of the wood further along the piece, causing the development of severe splits or seasoning checks. This defect can be partly avoided by the use of end cleats, generally pieces of wood or metal nailed across the end-grain of the boards in order to physically prevent the development of splits. This method of controlling checks is very unreliable and a more efficient technique is to seal the end-grain with tar, pitch or wax formulations so that drying is confined to the side grain throughout the length of each piece of wood. When the wood has a high moisture content in its initial green state it is comparatively flexible and it must be carefully stacked in order to avoid sagging. Stickers, or piling sticks, must be placed between the pieces to permit a proper circulation of air, or alternatively the pieces may be stacked with each alternate layer in a different direction, a method known as self piling which is frequently used for the drying of, for example, transmission poles. Sample boards must be incorporated within the centre of the stack so that they can be weighed at intervals in order to follow the drying progress. Where hardwoods are stacked in boule, they are sawn through-and-through and the boards later reassembled in their original order but with stickers between them to permit air circulation, see Figure 2.2.

Kiln seasoning

In kiln seasoning the wood is placed in containers or kilns in which the temperature and humidity can be controlled in order to achieve the maximum rate of drying consistent with freedom from the development of defects. The kiln must be designed and the wood carefully stacked so that a uniform air circulation can be achieved. If the air is always passing in the same direction it is obvious that the temperature will be lower and the relative humidity higher at the exit, so that the wood nearest the exit will have a higher moisture content than that close to the inlet. In many kilns the air flow can be reversed in order to reduce these differences. One advantage of kiln seasoning is the high temperatures that are used towards the end of the drying cycle as this eradicates any insect infestation as well as many fungal infections.

The high temperatures also mean that the kiln is expensive to operate unless care has been taken in providing adequate insulation. Most of the air is usually recirculated in order to conserve heat, a limited amount of air being vented to the atmosphere in order to remove the moisture. Kilns are not greatly affected by the relative humidity of the incoming air as it is usually heated and this immediately reduces the relative humidity. In fact, the air can become far too dry and steam jets are often used to increase its humidity in order to prevent too rapid drying which might cause damage.

Moisture in wood

Perhaps at this stage it would be sensible to discuss the relationship between the moisture content in wood and the atmospheric relative humidity. The ability of air to absorb moisture varies with the temperature so that the moisture content at which air becomes saturated also increases with the temperature. When air is completely saturated it is said to have a relative humidity of 100%, and a relative humidity of 50% naturally means that the air is half saturated. If air increases in temperature the relative humidity will fall, or the air will appear to become drier as it is capable of holding a larger amount of moisture at the higher temperature, yet the actual humidity or moisture content of the air will remain the same. Thus the drying power of air will be increased as the relative humidity is reduced, either by reducing the actual humidity or simply by increasing its temperature. Wood cannot dry if the relative humidity of the air is 100%. In addition if wood is allowed to remain in air possessing a constant relative humidity, it will eventually acquire a moisture content in equilibrium with the surrounding air. The moisture content in equilibrium with 100% relative humidity is known as the fibre saturation point and lower moisture contents are in equilibrium with lower atmospheric relative humidities.

The actual operating temperature of a kiln is comparatively unimportant as it is relative humidity that is responsible for the rate of drying. The relative humidity of the air can be increased by using steam jets or decreased by increasing its temperature, as previously mentioned. However, neither of these techniques can be used unless the relative humidity of the air can be regularly measured. The normal method of measurement consists of Wet and Dry bulb thermometers. The dry bulb thermometer records the normal temperature of the air whilst the wet bulb thermometer is equipped with a sleeve of porous material which is supplied with water which evaporates, cooling the bulb. If the relative humidity of the atmosphere is 100% the air is saturated so that the evaporation is prevented and the two thermometers read identical temperatures, but when the relative humidity is low the evaporation from the wet bulb is very high, cooling the bulb and giving a large difference in temperature between the two thermometers. This difference in temperature, related to the dry temperature, can be used as a method for estimating relative humidity. Human hair increases in length by about 3% over the complete relative humidity range and this change in dimension can be used to rotate a dial, giving a direct reading of relative humidity. Various electrical instruments have also been devised to measure relative humidity and many of these can be coupled through control units with shutters, steam sprays and heater controls in order to automatically maintain relative humidity in a kiln.

Relative humidity measurement

Compartment kilns

Compartment kilns are operated by natural convection currents. They consist of a compartment in which the wood is stacked, properly stickered in order to permit the unobstructed flow of air. Beneath the compartment is typically a duct through which air enters, passing steam jets and heating coils in order to control its relative humidity. At the top of the stack is a false ceiling with further heating coils above. Usually a duct permits the fresh air to enter the compartment through the floor on one side. The stack of wood is arranged across the compartment from floor to ceiling and from wall to wall so that the incoming air is forced to pass through the stack to the other side of the compartment where it is vented to the atmosphere through a flue. The problem with this system lies in the rather uneven drying, partly due to reliance upon natural convection and partly due to the fact that the stack is bound to be drier closer to the incoming air duct.

Fan-type kilns

Fan-type kilns largely overcome these difficulties as the air flow rate can be finely controlled. In some kilns centrifugal blowers feed air into the kiln but this system suffers from the disadvantage that the air flow rate from centrifugal fans is usually high, preferentially drying the portion of the stack to which it is directed. In addition, these centrifugal fan kilns generally have no facilities for recirculating the air so that they are comparatively expensive to operate. The most popular kilns are operated with internal bladed fans rotating fairly slowly, sufficiently to ensure good air distribution without excessive local concentrations of flow.

Figure 2.11 **Cross-shaft overhead fan kiln.** *(Crown copyright)*

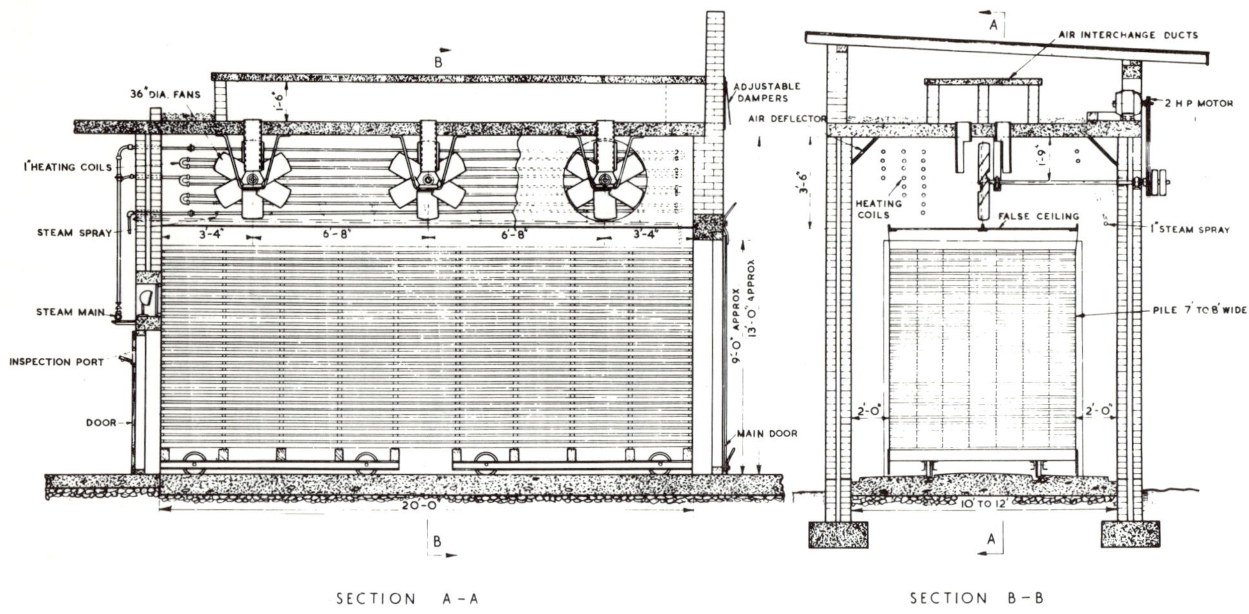

Progressive kilns

Kilns with internal fans can be very efficient but where there is a large throughput of the same species and sizes of wood it is possible to operate progressive kilning. In this system the stacks of wood are moved progressively through a long tunnel in which the air at high temperature and low relative humidity is introduced at the exit end, passing over wood that is already almost dry. As the air flows along the tunnel its temperature is reduced and its relative humidity increased so that it is almost saturated by the time it reaches the far end of the tunnel. The severity of the drying cycle depends upon the temperature and relative humidity of the air introduced at the exit where the dry wood is removed from the tunnel, coupled with the rate of flow of batches of wood down the tunnel.

Kiln schedules

Various kiln schedules have been devised to dry wood as rapidly as possible without the development of seasoning defects. Obviously the rate of drying will depend upon the cross-section of the pieces of wood and it is therefore important that a kiln should always be loaded with wood of similar cross-section. In addition kiln schedules are normally devised so that they are controlled by the moisture content of the wood at the air inlet side of the kiln. In a typical schedule it will be found that, when this moisture content reaches a certain level, the relative humidity of the air must be adjusted in order to proceed to the next stage. In many cases kiln schedules actually specify the wet and dry bulb temperatures that must be achieved at each stage as this is the most common method for monitoring the relative humidity within the kiln. In this way time is ignored in kiln schedules and it is thus unnecessary to define the actual sizes of the pieces of wood that are being dried. Different species of wood are able to tolerate different rates of drying. Thus Scots pine and abura are both able to tolerate very rapid or rapid drying and only about

$1\frac{1}{2}$ weeks is required to dry 1 in. (25 mm.) boards from green to a moisture content of about 12%. Beech must be dried less rapidly, taking $1\frac{1}{2}$ to $2\frac{1}{2}$ weeks to achieve the same result. Kokrodua must be dried more slowly, taking $2\frac{1}{2}$ to 4 weeks, whilst oak and greenheart must be dried very slowly, taking 4 weeks or more. The times taken for drying 2 in. (50 mm.) boards will be about two and a half times as long as those for 1 in. (25 mm.) boards.

Typical moisture contents

Fibre saturation point usually occurs at a moisture content of about 27 to 30%, and shrinkage occurs in wood dried below fibre saturation point. Air drying will normally reduce the moisture content to between 17 and 23%, and if lower moisture contents are required kiln drying is essential. Wood can be pressure treated with preservatives at moisture contents below 25%. Carcassing or framing timber in buildings can tolerate a moisture content of up to 23%, largely because its cross-section dimensions are relatively unimportant and drying is necessary only to achieve resistance to fungal decay. Wood intended for use in ships, boats and vehicles should be dried to 15%. In a house with reasonable central heating the average moisture content should be 12%, although bedroom furniture would perhaps be better manufactured with a moisture content of 14%, more closely related to the lower heating level in bedrooms. In buildings which are more intensively centrally heated, such as offices, a moisture content of 8% may be necessary and flooring installed on top of under floor heating should have a moisture content of only 6%. Unfortunately these various humidities are largely theoretical; it is very difficult to ensure that the wood remains completely protected during the numerous handling and transportation stages after leaving the kiln. The worst stage is perhaps installation in a new building where there is invariably a high moisture content which will automatically recondition the wood to a new moisture content.

Seasoning defects can still occur, despite great care in applying the most suitable drying schedules. Case hardening is perhaps the most common defect in certain species, such as beech. Where case hardening is anticipated or when it has actually occurred the kiln must be operated at a high temperature and high humidity in order to plasticise the outer layers of the wood so that the case hardening is relieved and the wood recovers its proper dimensions. This treatment is only applied for a very short period, sufficient to relieve the case hardening without significantly affecting the moisture content of the piece as a whole. High temperature is also used during kilning in order to ensure sterilisation of hardwoods against Lyctus beetle attack. In this connection it is interesting to note that kiln seasoned wood is far more susceptible to Lyctus than air seasoned wood, apparently because it possesses a higher starch content; in air seasoning the cells die only slowly after felling, significantly reducing the level of stored starch, whilst in kiln seasoning the cells are rapidly killed so that the starch remains. This is perhaps the one major difference between air and kiln seasoned wood and it is only of importance in the case of hardwoods which are susceptible to Powder Post beetles; see Chapter 3.

Mensuration

The measurement of wood presents some surprising difficulties. For many years the Imperial system of units has been most widely used but there is now a progressive move towards the universal adoption of the metric system. Unfortunately the Imperial and metric systems are both widely used at the present time, and in addition the adoption of the metric system has introduced some rather peculiar units of measure that are probably best described as "metric inches and feet"! These units apply particularly to sawn wood and panel products, and will be discussed in detail later in this chapter, but it is first logical to consider the problems involved in measuring logs.

Log volume

A perfect tree trunk is, of course, a steep sided cone and a log is therefore a frustrum of a cone with a true volume equal to the average cross-section area multiplied by the length. The cross-section area can be determined from a measurement of the radius r, the diameter d, or the girth g:

$$\text{Cross-section area} = \pi r^2 = \frac{\pi d^2}{4} = \frac{g^2}{4\pi} \qquad (\pi = 3.1416)$$

In fact "quarter-girth" is more often used than girth; the girth is measured using a piece of string which is then folded into four before being measured on a normal rule or tape, or a special tape is used in which the units are four times the normal length so that it gives a direct reading of quarter-girth. The reason for using quarter-girth rather than girth will be explained later. These systems for measuring the true volume of logs are used in the Customs Fund Measure system where the quarter-girth is measured by a tape or the diameter measured by calipers at the mid-point of the log and the following formulae applied to give the volume in Imperial units:

Customs Fund Measure

$$\text{Volume} = \frac{L(g/4)^2}{113} = \frac{L\ d^2}{183}$$

(volume in cubic feet, length L in feet, quarter-girth g/4 in inches, diameter d in inches).

The volume is said to have been calculated using the Customs Fund string or calliper measure, depending upon whether quarter-girth or diameter was employed.

The Customs Fund Measure assumes that the cross-section area at the mid-point of the log is the average cross-section for the length. In the Brereton system two measurements of diameter are made at each end of the log and these four separate measurements are then added and divided by four in order to obtain an average diameter. The results can be applied in the following formula:

Brereton system

$$\text{Volume} = \frac{0.7854\ L\ d^2}{144}$$

(volume in cubic feet, length L in feet, diameter d in inches).

The Brereton formula assumes that averaging the four diameters will yield an average cross-section area but this is not in fact correct; area is a square measurement and it would be necessary to average the squares of the diameters in order to obtain a true result, so that the Brereton formula results in a slight under-measurement. A further point is that the diameters are generally determined under the bark in order to measure the actual wood content instead of the gross volume of the log, as determined by the Customs Fund Measure.

In order to clarify the system of measurement it is usual to mark volumes "O.B." if they are measured over the bark but it will be appreciated that even under bark measurements do not necessarily indicate the volume of usable wood that is contained within the log. The Hoppus string measure was originally introduced in the early eighteenth century in an attempt to estimate usable volume. In the case of a log a mean quarter-girth would be determined

Hoppus measure

by measuring at the mid-point and the following formula applied:

$$\text{Volume} = (g/4)^2 \times L/144$$

(volume in cubic feet, length L in feet, quarter-girth g/4 in inches).

The Hoppus string measure is often known as the quarter-girth method as it originates from the concept that the usable volume of a log is approximately equal to a square baulk which can be cut from it, and the sides of this baulk are approximately equal to one quarter of the girth of the log. In fact, although this method of calculation is now known as Hoppus measure, the Hoppus Calculator actually consists of a system of tables designed to simplify the calculation of volume of all types of solid materials. Thus, in table 1, quarter-girth or diameter of a log is converted to cross-section which is expressed in terms of the length of the side of a square of the same area. In table 2 this square section is combined with the length to give the true volume in cubic feet. The process of using diameter or quarter-girth in table 1 in order to calculate the equivalent square area, followed by reference to table 2 to obtain volume is rather tedious and for many years the timber trade has tended to ignore the corrections in table 1 and to apply quarter-girth directly in table 2. Whilst this simplifies the measurement procedure it also introduces a deliberate underestimate of 21.5% compared with the true volume. In fact, Hoppus measure is also used to estimate the volume of a standing crop of trees based upon measurements of the quarter-girth at breast height coupled usually with an estimate of the total height of the tree, but this is a subject that is more appropriate to a book on forestry.

Softwood sizes

It might be imagined that the measurement of sawn wood would be rather simpler than for tapering logs with or without their bark. In fact there is perhaps even more confusion because of the various terms and units which are commonly used. Softwood from the northern coniferous forests is generally sawn into blanks or boards about 2 to 4 inches thick, which are described as deals (9 to 11 inches wide), battens (5 to 8 inches wide) and scantlings (2 to 4½ inches wide). Pieces less than 2 inches thick are known as boards (over 4 inches wide) or strips (less than 4 inches wide). Pieces with large cross-sections, 4 by 12 inches and upwards, are known as flitches, whilst large squared logs, 9 by 9 inches, 10 by 10 inches, 12 by 12 inches, etc., are known as timbers. Partly squared logs with waney edges are known as baulks. In all cases a conversion factor of 25 mm. to 1 inch gives the equivalent metric dimensions.

Standards

When shipped in bulk wood is normally measured in Standards, or Standard Hundreds (S.H.). There are a number of hundreds originating from different exporting ports but the most important is the Petrograd Standard which is now used almost universally and referred to briefly as "P.S.H.", or simply a Standard. This consists of 165 cubic feet and was originally derived from 120 pieces, each 12 feet long, 11 inches wide and 1½ inches thick. The only other important Standard is the Gothenburg Scale Standard which is used for round wood, particularly pit props.

Hardwood sizes

Hardwood is generally flat sawn, through-and-through, and is described as a blank if more than 2 inches thick and a board if less than this thickness. After resawing to give a square edge it is still described as blank if more than 2 inches thick but narrower thicknesses are scantlings, further divided into boards which are more than 4 inches wide and strips if less than this width. The measurement of hardwood can present some difficulty because of the peculiar manner in which the results are expressed when the Imperial system is used. If length is in feet but width and thickness in inches, the following formula should be used to find the volume in cubic feet:

$$\text{Volume} = L \times W/12 \times T/12.$$

In fact quantities are often expressed in cubic feet, inches and parts, an inch being 1/12th of a cubic foot and a part 1/12th of an inch so that these are the remainders from each division by 12. Thus for a piece 20 feet by 4 inches by 8 inches the multiplication of these figures will give 640. The first division by 12 gives 53, remainder 4 inches. The second division by 12 gives 4 cubic feet, remainder 5 parts. This system is also sometimes used for softwood.

Board measure

In North America and Australia wood is often measured in feet board measure (Ft.B.M.), or thousand feet board measure Mft. B.M.). One board, or superficial, foot is 1 foot run of 1 inch by 12 inch board, and thus equals 1/12th cubic foot. For boards more than 1 inch in thickness this relationship always holds true so that one board or superficial foot is always 1/12th cubic foot, but for boards less than 1 inch thick a board foot is based upon the superficial area only by multiplying the lengths in feet by the width in inches. The Brereton and Hoppus systems are often used to express the contents of logs in board feet simply by replacing 144 by 12 in the previously quoted formulae.

Load

There are a number of other less important units of measurement, such as a "Load" which consists of 50 cubic feet of sawn wood or 40 cubic feet of roundwood or a Hoppus Load which consists of 50 Hoppus feet of roundwood. The ton is also used for some wood but this is generally derived from volume measure; for air-dry softwood a Standard is assumed to weigh 2.5 tons for pieces up to 4 inches thick or 3.12 tons for pieces over 4 inches thick.

Actual and nominal dimensions

Finally, resawing and planing reduce pieces of wood in size. The final dimensions are "actual" whilst the original size is "nominal". Normally resawn or planed pieces are referred to by their nominal dimensions, so that planed square edge (P.S.E.) boards are about $\frac{1}{4}$ inch less than their nominal width. In the case of planed tongued and grooved (P.T.G.) boards there is a further loss of dimension due to the tongue and groove, so that 1 by 6 inch nominal boards are $\frac{7}{8}$ by $5\frac{3}{4}$ actual but $\frac{7}{8}$ by $5\frac{1}{2}$ laid; there is a further loss of $\frac{1}{4}$ inch when the tongue fits into the groove on the adjacent board.

Metrication

The universal introduction of the metric system would appear to considerably simplify these measurement problems by introducing a series of consistent decimal units. It is therefore to be greatly regretted that the trade has chosen generally to ignore the decimal advantages of the metric system and instead to introduce a special new system of units which might be best described as metric inches and metric feet. It will be appreciated that there is no real advantage in adopting metric units of measurement in place of Imperial units unless designers also adopt the metric module at the same time. The whole point of metrication is decimalisation and the metric modules must therefore be decimal in concept to gain the maximum advantages from the system. The metric system as generally adopted involves the use of millimetres as the unit for cross-section measurement, and generally the cross-section will vary in stages of 25 mm. This 25 mm. unit has been adopted as it is close to an inch (25.4 mm.), and as it is also a quarter of 100 mm. it is reasonably decimal in concept. Unfortunately the decimalisation fails for thicknesses below 75 mm. where the trade has unfortunately adopted 63, 38, 32 and 19 mm. as standard thicknesses, simply converting direct from the old Imperial measure. This is not, of course, metrication and the industry should have adopted perhaps 65, 40, 30 and 20 mm. in order to preserve the decimal concept of metrication. In length the situation is perhaps even more ridiculous. Although it is said that lengths must be measured in metres, the trade has adopted 300 mm. as its standard cutting unit as this is very similar to the Imperial foot. For example, a piece of wood is now available which is 3.6 metres long, or 12 units of 300 mm. Unfortunately, a design in metric measurement will certainly be decimal in concept so that the required length is likely to be 3.5 metres and 0.1 metre will be wasted. Alternatively, a design in Imperial units might involve a length of 12 feet but this piece of 12 x 300 mm. units is only 11 ft. 9¾ in. and cannot be used. The next size is 3.9 metres so that about 9½ inches must be cut to waste. Despite these peculiar lengths and cross-section units it is still considered necessary to buy and sell wood in square metres or cubic metres, yet these units make it very difficult to calculate area and volume. There is now a considerable danger that, without a rational decimal system, certain sections of the trade will progressively ignore this metric system and adopt instead some new and arbitrary system involving simpler calculations, perhaps rather like the old Board feet or Standard Hundreds system.

This rather peculiar metrication has been extended to plywood, block board, particle board and fibre board. These panel materials must be bought and sold in units of 10 square metres, yet again no attempt has been made to rationalise their measurements on a decimal system. Instead the boards continue to be manufactured to the old Imperial units of measure and the closest metric conversion adopted. Thus a board which is 4 by 8 feet becomes 1,220 by 2,440 mm., and if it is ⅜ inch thick it becomes 9.5 mm. It is rather difficult to understand that these are actually standard recommended dimensions and it will be clear that the introduction of metrication has resulted in enormous complications instead of the simplification through decimalisation that was originally intended.

2.2 Woodworking

The use of various saws for the initial conversion of logs into sawn wood has already been mentioned. Only a limited depth of cut is possible with circular or rotary saws so that band-saws are more often used where large diameter logs are converted. In other mills reciprocating verticle frame or gang-saws are employed to give multiple cuts when a log is being cut through-and-through or flat sawn. In general band-saws can be used for almost any conversion purpose whereas frame or gang-saws must be rebuilt whenever a different thickness of board is required. Circular saws can only achieve limited depth of cut and, when larger diameter saws are required, the gauge must be increased to give the necessary stiffness, giving a wider kerf or saw cut so that more power is required and there is a higher wastage.

Saws

In woodworking the same equipment is used, although it is often smaller in size as it is used only to reduce wide boards or large flitches to the required commercial sizes. Whilst a circular saw with an adjustable fence is a particularly adaptable piece of equipment, it is generally fed by hand and reciprocating gang or frame-saws are preferred wherever there is a large amount of repetitive sawing to produce particularly popular sizes.

It is necessary to use a particular terminology when describing saws. The thickness of the blade is termed the gauge, whilst pitch refers to the distance between the points of consecutive teeth. The space between the teeth is the gullet and the depth is the distance from the deepest point of the gullet to a line joining consecutive points. The hook angle is measured between the face of the tooth at the point and a line perpendicular to the line of points, or a radius line in the case of a circular saw. The back angle is measured in the same way but refers to the angle of the back of the tooth at its point. The set of a saw is the projection of the teeth on either side in order to provide clearance, spring set involving the springing of the teeth on alternate sides and swage set involving inserted teeth, each wider than the gauge of the blade so that they project equally on either side. In the case of a circular saw the centre is usually tensioned or expanded so that the cutting edge remains taut, even when it is heated during working.

Figure 2.12 **Parts of a saw blade.**

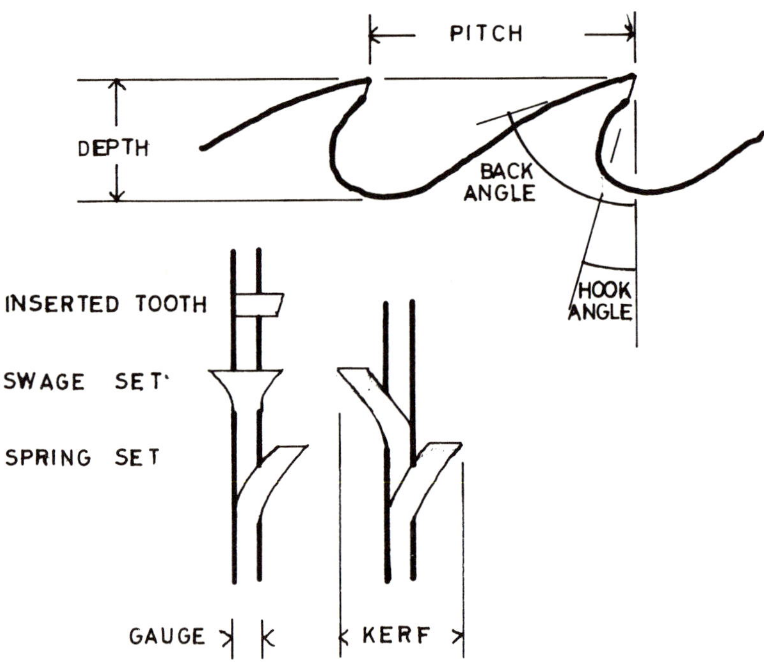

Circular or rotary saws

Generally a circular saw should be the minimum diameter required to obtain the necessary cut as larger diameter saws are necessarily thicker gauge and involve a wider saw kerf and a greater power demand. The necessary gauge for a particular diameter would depend upon the type of wood to be sawn, the depth of cut, the rate of feed and the power available, but it is essential that a saw should be sufficiently stiff to avoid flutter and wander, and heavier gauge saws are required for hardwood than for softwood. The speed of the saw is adjusted normally to give a tooth speed through the wood of about 10,000 ft. per minute (3,000 m. per minute) so that a 30 in. (75 cm.) diameter saw would run at about 1,280 revolutions per minute but a 60 in. (150 cm.) diameter saw at only 640. The number of teeth on a circular saw will depend upon the rim speed, the wood to be sawn and the rate of feed but generally 54 to 66 teeth are used on saws of all diameters; more teeth are only desirable when cross-grained or knotty wood is being sawn or when the saw is used for cross-cutting. It is true that a greater number of teeth will give a smoother cut but a heavier gauge is then necessary and more power will be required. Indeed, an excessive number of teeth will simply result in the sawdust being cut into a finer powder, involving overheating and the use of excessive power. Indeed it is pitch, or the distance between the teeth, that has the greatest influence on power requirements and it is generally agreed that economy is best achieved by using the minimum number of teeth possible to give the desired cutting action. A reduced number of teeth on a saw involves careful preparation as they must each do more work, yet it is often found that teeth will lose their sharpness more rapidly if they do not make a clean cut and too many teeth on a saw will often result in rapid blunting.

Figure 2.13 Circular saw teeth and cutting patterns.

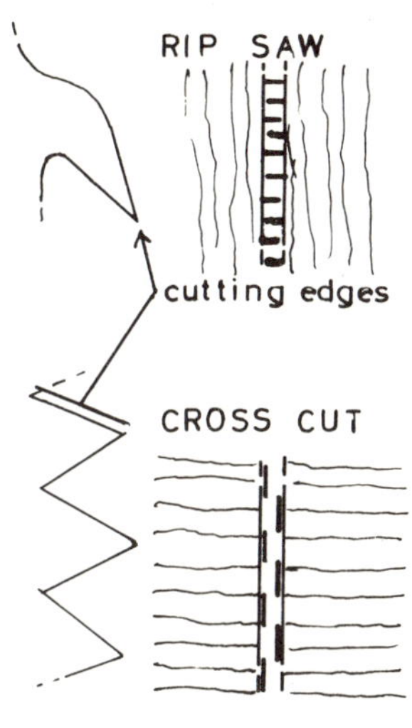

There are many conflicting views on the most desirable tooth shapes for particular purposes. However, in general principle, it may be said that the tooth must cut across the grain so that, in cross-cutting, the side of the tooth must do the work whilst in rip sawing along the grain the leading edge must be a chisel led to cut across the grain. However, in rip sawing the chisel edge alone is often found to be inadequate, particularly in hardwoods where it can lead to overheating, and some cutting action by the outer edge of the tooth is necessary in order to allow for diagonal or distorted grain. In the case of a rip saw a greater hook will give a finer cut as the chisel edge will be cutting perpendicularly across the fibres or tracheids so that a saw with greater hook will cut more easily and use less power. However, a finer tooth is also more susceptible to wear or damage when it encounters a harder zone of wood such as a knot. Generally a hook angle of about 30 degrees is used for circular saws for all woods. The gullet space must not be too great as it will also lead to weakening of the tooth but it must be sufficiently large to carry away the sawdust as clogged gullets can lead to serious overheating.

The set of a saw, the projection of a tooth beyond the blade, must be adequate to provide proper clearance but it must not be too great as an excessive amount of wood will then be removed, requiring additional power. For hardwoods a set of about 0.012 in (0.3 mm.) is required but this should be increased to about 0.018 (0.45 mm.) for softwoods which are more elastic and have a greater tendency to jam the blade.

The saw speed will depend partly upon the feed speed, although generally the tooth speed must be maintained at about 10,000 ft. per minute (3,000 m. per minute). For initial conversion the feed speed and saw speed should be adjusted to give a feed per tooth of between 0.05 to 0.125 in. (1.2 to 3.0 mm.), although there is a possibility that a particular saw may be unstable when run at these speeds and slower feed rates may be desirable.

Band-saws

Band-saws vary in dimensions from about 6 in. (15 cm.) to about 14 in. (35 cm.) in width, the length of the blade varying according to the wheel diameters and the depth of cut required. In principle tooth design is the same as for circular saws, except that the saw always has the same angle of attack to the wood so that band-saws can be more critically designed and are generally more economic than circular saws. Saws used for converting softwoods must be designed firstly with ample gullet space and must also have a greater hook angle than those used for cutting hardwoods so that the leading chisel edge will cut cleanly through the tracheids without causing compression and loss of efficiency. Saws used for cutting hardwoods require very strong teeth and this necessarily results in a reduction in the hook angle and the gullet space, reducing the rate of feed that can be handled by the saw. Band-saw teeth are usually swage-set by squeezing the tooth points with a "swage" in order to obtain the necessary projection on either side; a swage can also be used for setting circular saw teeth, although inserted tungsten carbide teeth are more normally used. The saw blade is tensioned by expanding the centre of the blade using special powered rollers so that the outer edges are under tension when the saw is stretched on the wheels. If tensioning is properly applied the back of the saw is expanded or stretched relative to the toothed edge so that the squareness of the saw is retained as the cutting edge expands due to heating. The true running of the saw through the wood is achieved by running the blade through guides immediately below the saw table and on an adjustable arm above the wood; this arm must be kept as close to the wood as possible in order to avoid vibration in the blade and a tendency for the saw to wander as it encounters knots or cross-grain. Band-saws generally run at about the same tooth speed as circular saws, although lower speeds must be used in some circumstances, such as when a strong tooth is necessary for cutting hardwoods and the gullet space is limited as a result.

Gang or frame-saws

The gang-saw, consisting of a number of saw blades mounted in a verticle frame, operates with an orbital reciprocating motion so that it cuts only on the downward stroke. The tooth design is basically similar to that adopted for band-saws. A gang-saw has the advantage that, as it is making a number of cuts through the log in a single operation, it can achieve a high rate of production with comparatively limited handling. It is also simple to operate and relatively low in maintenance cost, achieving the minimum of waste in sawing and great accuracy which makes it particularly suitable for preparing thin boards. The disadvantages of a gang-saw are its relatively high initial cost and its inflexibility; considerable time is required to reset the saw to cut different thicknesses. In addition the wear is concentrated over a limited length of the blade, whereas it is evenly spread in circular and band-saws.

Planing

Normal sawing leaves wood with a rough surface which has been torn by the saw teeth. Occasionally special saw blades are used, usually known as splitting saws, which have a sharp edge to each tooth which smoothes the cut but generally sawn wood must be planed to obtain a smooth surface. In hand planing a sharp blade is passed across the surface of the wood in order to remove a thin shaving. In mechanical planing the blade or cutter is mounted in a cutter block or head which is rotated at high speed on a spindle above the surface of the wood so that, as the wood passes beneath, each blade will remove a short shaving. A machine which planes or dresses the wood to a smooth surface is known as a surfacer or planer, or thicknesser if both faces are planed in one operation. In fact machines are often designed to plane on four sides in a single operation and the blades may be shaped in order to achieve various mouldings, the machine then being known as a moulding machine, a four header or a four cutter. In many cases additional cutters are provided, usually up to seven, in order to achieve special mouldings or perhaps to split the moulded material as it emerges from the machine in order to double the rate of production.

Figure 2.14 Cutters for planing and moulding.

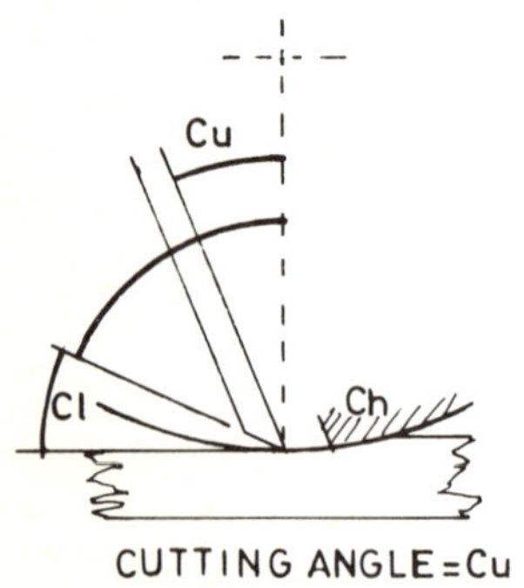

The cutter blocks rotate at very high speeds, usually about 3,500 r.p.m., and they must be very finely balanced in order to completely eliminate vibration. The cutting or approach angle is the angle between the cutting edge of the blade and a radius through the cutter spindle. Obviously a large cutting angle will give more efficient cutting but in woods with irregular grain, particularly interlocking grain when quarter sawn, a large cutting angle will tend to result in grain raising and roughness. Conversely a greater cutting angle is required for wood at high moisture content as there is otherwise a tendency for a blade with a low cutting angle to compress the wood elements rather than cutting so that they later recover their original dimension and result in raised grain. The cutting angle is determined primarily by the design of the cutter block and the diameter of the cutting circle but the cutting angle can also be reduced by honing the leading or cutting edge of the blade.

Obviously cutting angle is only one of the several factors that affect the finish, and the number of cuts is perhaps almost as important. Mechanical planing results in a series of small waves on the planed surface so that increasing the number of cuts, perhaps by reducing the feed rate, makes this wave pattern less apparent. To take a number of examples, unseasoned western red cedar requires a cutting angle of about 30° and perhaps 8 cuts per inch (3 cuts per cm.), but when seasoned the cutting angle should be reduced to about 25° and the cuts increased to about 10 per inch (4 per cm.). The cutting angle must be reduced and the number of cuts increased progressively for Baltic redwood (Scots pine), Radiata pine, Japanese oak and Queensland maple, the latter requiring a cutting angle of 15° and 12 cuts per inch (5 cuts per cm.) unseasoned and 10° with 15 cuts per inch (6 cuts per cm.) when seasoned.

The back of the cutter knife is bevelled to give a clearance angle and also to give a sharper but weaker cutting edge to the blade. A certain minimum clearance angle is necessary to prevent the back of the blade from jamming against the surface of the wood, and this angle will be less for cutters of larger diameter. It will be appreciated that each cut ends as the knife passes upwards and breaks through the surface of the wood so that there is a tendency for the wood to be lifted by the knife. This lifting is avoided by the chip breaker, or a continuation of the cutter block which tends to press down on the wood at this point and which checks a tendency for chips to break out of the wood in advance of the knife.

Drills

Drills are simply small cutters rotating about a spindle held perpendicular to the surface of the wood. The high speed twist drill is perhaps the most simple system in which a spiral is cut in the drill shank, and, at the point of the drill, the edge of the spiral is sharpened to provide a cutting blade. In addition the point of the drill is bevelled so that it can be readily located on the wood in the position in which a hole is required. The traditional woodworking bit is, in contrast, a very low speed device consisting of a pin to locate the bit on the wood and two wings. One wing is fitted with a scribing knife which first cuts into the wood to mark out the precise circumference of the hole and the other wing is fitted with a horizontal cutting blade which removes the wood from the interior of the scribed hole. These two types of drill represent the basic principles and there are many hybrid devices that combine scribing knives to cut out the circumference, horizontal knives to remove the volume of the wood and spiral fluting to provide a means for removing the debris from a deep hole. An auger is perhaps the best example of a combined drill which is used at very low speed but which possesses all these features.

2.3 Wood based products

Wood is, of course, a natural material produced as the trunk of a tree and this necessarily results in some restrictions in its form and dimensions. The wood in the trunk is naturally in the form of a steep sided cone so that length can be obtained fairly readily but width is limited by the diameter of the trunk. In fact, lengths are also irregular as the sawing of a trunk from the bole towards the crown will necessarily result in the outer pieces having an increasing amount of wane until the saws eventually emerge from the log, so that only the central pieces will possess the full length of the log. Random lengths are difficult to handle for transportation and there is a tendency for long pieces to be trimmed to a suitable size for strapping into rectangular packages. This necessarily means the wastage of both the trimmings and also the shorter pieces obtained during the initial conversion of the log.

The logical way to avoid this wastage is to join all pieces leaving the conversion saw end-to-end to form one continuous length of wood which is then cut to the required lengths for strapping into packages. The simplest method for joining in lengths is the butt joint in which the ends of the two adjacent pieces are carefully trimmed, spread with adhesive and forced together. This system has the advantage that the waste is negligible but, even with a perfect adhesive, a butt joint possesses very low strength because of the low contact area that results when you attempt to join end grain with end grain; the simplest comparison is to envisage the problem of joining two bundles of random sized tubes together. As a result it is difficult to develop a joint which is more than 10% efficient, or which has more than 10% of the strength of the unjointed wood on either side. Much more efficient adhesion can be obtained by glueing side- or edge-grain to produce a lap joint, but in all cases lap joints involve greater wastage as the two adjacent pieces are required to overlap before trimming to form the joint. It might appear that the perfect joint would be a comb joint in which the ends of the adjacent pieces are slotted and then slid together so that there is a large area of contacting side grain to provide an efficient contact surface for adhesive. In fact, the tip of each tongue represents an inefficient butt joint so that the total efficiency is unlikely to exceed 50%. A comb joint has the advantage that it can be relatively short with only limited wood wastage but it is very difficult to assemble in practice; it must be accurately machined if a sufficiently tight joint is to be obtained to give high strength, yet a tight joint wipes off the adhesive during assembly and, in fact, results in a relatively weak joint.

Butt joints

Comb joints

Scarf joints

In a scarf joint the ends of adjacent pieces are shaped to form sloping planes which have a close approximation to side grain and thus give reasonably good strength efficiency. In fact, normal slopes of 1 : 8 to 1 : 12 will give theoretical strength up to 95% but practical assembly problems reduce this effectively to about 85%. The slope of the grain is very significant and it is also difficult to cut a perfect feather edge as there is a tendency to produce a slightly concave cut. In addition it is difficult to register the sloping surfaces during assembly; lateral clamp pressure or end pressure are both likely to cause slip. Nails are sometimes used to ensure close registration of the contacting surfaces but this necessarily involves an additional assembly operation. A scarf joint can be modified to incorporate a step at the end or middle of the slope in order ensure correct registration but this seriously complicates the cutting operation. Finally the strength of the joint falls considerably if the slope is less than 1 : 8 and, although it increases little beyond 1 : 12, this still means a long and wasteful joint.

Figure 2.15 Slope and scarf joint strength.

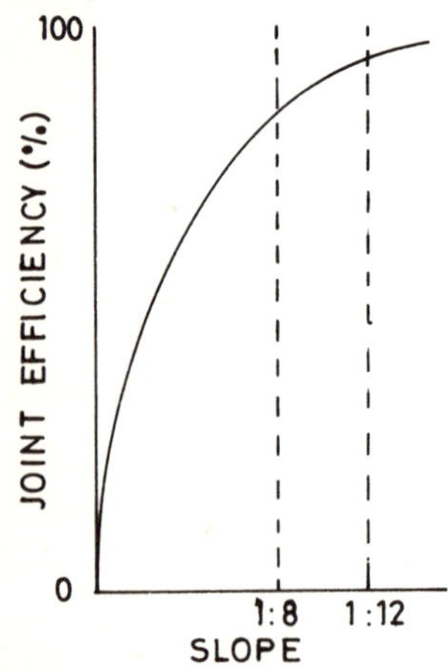

Finger joints

Finger jointing was first developed by Dr. Egner of the Otto-Graf Institut, Technische Hochschule, Stuttgart during World War II in an attempt to improve the utilisation of wood during the war time shortages. A finger joint consists of a comb joint in which the fingers are tapered so that the final joint is a multiple scarf joint. In this way all the advantages of both the comb and the scarf joints are combined without their attendant disadvantages. The

Figure 2.16 **End-to-end joints.**

short comb format of the joint ensures low wastage whilst the scarfes give high adhesive efficiency, perhaps as high as 96% of that of clear unjointed wood. In fact it can be considered in practice that the efficiency of a finger joint will exceed 80% and the defect that it introduces within the length of a piece of wood is less than that arising through the presence of even small knots.

Many different finger joint profiles have been proposed in order to improve the degree of side grain adhesion or, in the case of a staggered profile, in order to reduce stress concentration. Recent developments have included a mini joint with very short length and consequently low wood wastage, and the even smaller die formed joints which reduce wastage to an absolute minimum and which also simplify the production processes.

Generally the joint profile or plan appears on the wide face of a board. A completed finger joint consists of a series of self-locating scarf joints but typically with very small butt joints at the tips of the fingers. These butt joints represent a weakness and must be kept to a minimum but they have a very real value. Pointed cutters will overheat and wear quickly so that the base of a V-cut will not be sharply defined and the tip of the corresponding finger will not be forced fully home during assembly, giving a loose and weak joint. In practice a small gap is therefore designed between the end of each finger and the base of its corresponding V section.

It has already been explained that long cuts for scarf joints tend to give a concave surface and long finger joints are impracticable for the same reason, yet a shallow slope is essential to give strength and this necessarily results in a long scarf. In finger jointing a compromise is normally reached by using a length to pitch ration of 4 : 1, equivalent to a scarf slope of 1 : 8. Normal size finger joints involve length/pitch ratios of 50 mm./12 mm., 20 mm./6.2 mm. and 10 mm./3.7 mm. In smaller profiles the number of fingers is increased in order to preserve the preferred length/pitch ratio, although in the smallest profiles a lower length/pitch ratio is found to be acceptable; it theoretically gives a weaker joint but it is found in practice that this is more than compensated by the ease with which the joint can be assembled to give a tight adhesive line. In fact, end thrust during assembly is very important if full strength is to be developed; a minimum of 200 lbs./in.^{2} (1,400 kN/m.^{2}) of cross-section is required for large profile finger joints in softwoods and not less than 300 lbs./in.^{2} (2,100 kN/m.^{2}) for denser hardwoods. These pressures are not theoretically necessary to ensure good contact of the surfaces to which the glue has been applied but are instead required to ensure that the joint remains assembled after the end thrust pressure has been released; the adhesive will cure only slowly, long after the piece of wood has left the assembly machine.

Plate 2.10

Finger jointing machine with, to the right, a high frequency curing unit to ensure that the adhesive has developed full strength before the wood is cross-cut to length. *(Cook Bolinders Ltd.)*

The procedure is to feed random lengths of wood into a machine which cuts profiles in the ends on either side of the joint. Adhesive is applied to the profiled ends and the pieces rammed together. In practice higher pressures are used than those previously quoted as the minimum, typically about 450 lbs./in.^{2} (about 3,100 kN/m.^{2}) for softwoods. It should be appreciated that this represents a considerable end thrust, about 2.4 tons in the case of a piece of 2 by 6 in. (50 by 150 mm.) cross-section so that difficulties are encountered in clamping the pieces. There is a danger that crushing will occur or, if the pieces are cupped, splitting is probable if there is an inadequate clamp area. In small profile finger joints with lower length/pitch ratios even higher end thrusts are required, perhaps 1,000 to 2,000 lbs./in.^{2} (7,000 to 14,000 kN/m.^{2}) and this requires an even longer clamp area in order to spread the load and avoid crushing.

As an alternative to cutting it is possible to impress the profile on the end of the piece of wood using a shaped die. The wood is first cut square and then rammed against the die, pressures of 5,000 lbs./in.^{2} (35,000 kN/m.^{2}) being required for making the die impression and perhaps half this pressure during assembly. Because of the obvious technical problems the die impressed profile method is not yet widely used for finger jointing but it is suggested that it is most suitable for wood which is plasticised to a certain extent by a high moisture content so that the use of a hot die will readily penetrate the wood and then set it to the required profile. Where fingers are cut or impressed using a die it is important that the moisture content should be appropriate for the type of adhesive that is employed, and particularly important that the moisture content of the pieces on either side of the joint should be within 3% so that subsequent drying or wetting does not cause distortion.

Radio frequency curing can be used if the joint is required to have early strength, such as when the wood is to be worked immediately after assembly. Finger jointing is particularly efficient and introduces less weakening in a piece of wood than the common defects, such as sloping grain or knots. The use of finger joints is particularly highly developed in Scandinavia where industrialised buildings and joinery (millwork) components are now widely manufactured from finger jointed wood.

Lamination

The lamination of thin pieces of wood to form a thicker piece has several advantages. In a single piece of solid wood a defect, such as a knot, is concentrated at the point at which it occurs but if the wood is cut into thin sections and mixed in a random way before assembly into a laminated piece

Plate 2.11 **Laminated beams under construction.**

the concentration of the defect is then lost and the point of weakness removed. Lamination can therefore be used as a means for producing continuous pieces of wood of very high strength and virtually free from defect or, alternatively, lamination may be used as a means for utilising low grade wood for relatively high grade purposes; it has been explained earlier in this chapter that the grading requirements for laminated wood are less rigorous than those for solid wood. The use of laminated components will be explained more fully in the wood engineering section of Chapter 4.

Plywood

In plywood very thin sections, veneers, are employed to produce panels.

Plate 2.12 **Plywood production. Peeling or rotary cutting plywood veneers.** *(UAC Timber)*

Plate 2.13 **Plywood production. Adhesive coated veneers leaving the spreader.** *(UAC Timber)*

Plate 2.14 **Plywood production. Adhesive coated veneers being fed into the press after assembly as a sandwich.** *(UAC Timber)*

Any defects in wood stock are scattered randomly throughout the panels so that their influence is virtually insignificant, as in laminated beams. In addition, adjacent veneers are orientated at right angles so that a joint at the edge of one veneer is bridged by the length of the neighbouring veneer. This also means that movement in the cross-section dimension of one veneer is restrained by the stable longitudinal dimension in the adjacent veneer, so that plywood is a method for utilising wood for production of strong, continuous and stable panels. The veneers are produced by sawing or slicing, or by the rotary peeling of logs; the latter method is used almost exclusively for structural plywoods. Although some decorative veneers can be produced by rotary cutting, particularly the flame figure veneers that are commonly produced from Douglas fir and other softwoods, the true figure of other decorative woods can only be produced by slicing. For example, the silver figure in oak and the stripe figure in tropical hardwoods with inter-locking grain can only be developed on radial or quarter surfaces. These veneers are therefore expensive and are generally supported on other less valuable veneers prepared by rotary peeling, so that the manufacture of plywood represents a particularly efficient means for the economical utilisation of rare and beautiful woods. Naturally these decorative veneers do not necessarily need to be applied to plywood but can also be used on particle board or solid wood; the main requirement is that the backing should be stable in the cross grain direction of the veneer, so that the veneer should normally be orientated at right angles to the grain of a solid wood support.

Figure 2.17 Various plywoods.

Veneers are normally cut from wood at high moisture content so that, after trimming to size, the veneers are dried before the application of adhesive and assembly. Hot pressing is used to ensure good adhesive contact and in order to achieve rapid cure. The resulting product represents wood with its natural advantages or disadvantages orientated in two directions instead of one as in the natural tree. However, it is occasionally advantageous for plywood to have higher strength in one direction than another and this is normally achieved by the manufacture of block board, batten board and laminboard. In blockboard a series of small square section sticks of wood are assembled side by side to form a sheet and one or two veneers added on either side. Battenboards are assembled in exactly the same way except that wider section sticks are used and, in laminboard, very thin sticks of wood are employed.

Particle board

One of the principal advantages of plywood is its low movement with change of moisture content. This is achieved by orientating adjacent veneers at right angles, so that the unstable cross-section of one veneer is restrained by the stable longitudinal section of the adjacent veneer. This enables plywood to be used for large panels. An alternative method for producing panels consists of preparing particles or chips which are then randomly mixed with adhesive and compressed to form a board. In particle board the movement is randomly orientated in all directions, although the large adhesive concentration tends to give some restraint. This method is not as efficient as plywood in controlling movement and the surface chips in particular are susceptible to an increase in thickness which may cause shadowing or unevenness through surface veneers or finishes. In addition particle board is much weaker than plywood, as the adhesive joints between the individual chips involve end grain surfaces and thus the inefficient bonding previously described for butt joints. However, particle board can be prepared from very low grade wood, small diameter logs derived from thinnings or the tops of trees, and species which, because of their physical properties as solid wood, are not normally utilised for constructional purposes.

Assuming that an appropriate adhesive has been employed, the properties of a plywood depend largely upon the wood species that has been used. Whilst the properties of a particle board are, of course, largely dependent upon the adhesive that is employed, the particle shape is of very considerable significance. For example, if the particles are all cubes the formation of the board will necessarily result in a large proportion of joints involving end grain, thus producing a weak board. In contrast long thin chips will tend to form a "felt" in which the individual chips overlap rather than butt, so that the joints are largely side grain and a very strong board results. Whilst high pressures are invariably used in the formation of the boards there is naturally a tendency for compressed particles to recover their dimension when the pressure is released and, to use the previous extreme example, the use of cube particles will necessarily result in a lumpy surface. The use of large chips, particularly if they are long and flat so that a substantial overlap can be obtained, will result in a high strength board, yet the surface texture will be rather coarse and may not be suitable for the application of a finishing system, such as paint to wall panels or varnish when the board is used for flooring. In order to avoid this problem boards are frequently manufactured in three layers, the centre layer involving large long chips to ensure high strength whilst the outer layers consist of small particles to give a smooth surface.

Whilst particle boards are generally manufactured from wood, a variety of other materials can be employed, particularly flax shives, or the woody core of the flax plants used for the production of linen. In the case of logs, the first stage is soaking and debarking, followed by cutting the logs into suitable lengths for feeding to the chippers; generally these involve the logs being abraded away by knives fixed on shafts rotating at high speed. The wet chips produced in this way are passed through hammer mills which compact them and reduce them to their final dimension; at this point other materials, such as raw planer shavings or plywood trim may also be introduced. After passing through a drier to control the moisture content the chips pass to a storage bin and then a batch weigher before being mixed with the adhesive and passed to a further storage bin.

The adhesive coated chips are then scattered on plates to form a mat. In some plants a mat may be formed individually for each board but in other plants the chips may be spread progressively in an endless belt. This is the stage at which multi-layer boards are formed. In some plants two particle lines are employed, so that firstly a layer of fine particles is deposited, followed by the coarse particles and a final deposit of fine particles to give a three layer board. Various proprietary plants have been devised to form these boards more simply. For example, in the BAEHRE-Bison forming machine particles of mixed size are introduced at a point over a moving endless belt and they are then blown by a strong current of air in both directions along the belt. The fine particles are blown the greatest distance so that, as the draught is blowing both ways along the belt, they are deposited both first and last, giving fine particles on both sides of the finished board. In the case of continuous belt manufacture the mats are then sawn into board lengths before initial pressing to compact the chips. This is followed by a final hot press to cure the adhesive and to give the board the designed density. Some distortion of the board will occur during hot pressing and further trimming is necessary. At this stage the board has a very low moisture content and it must be transferred to a maturing area to allow it to regain some moisture content, thus allowing the chips to reach their equilibrium dimension before the board is sanded to a smooth surface.

Whilst this method of manufacture is normally used there are a number of other techniques of minor importance. In the vertical and horizontal extrusion processes the chips are not scattered on a fixed or moving platen but instead they are forced between two parallel heated metal plates by a reciprocating ram so that a board is extruded. These extruded boards have rather different properties than flat pressed boards as the particles tend to be orientated perpendicularly to the plane of the board. Thus extruded boards tend to have excellent dimensional stability in thickness but low stability in the plane of the board. In addition the boards have lower strength and a relatively coarse and porous surface. The manufacture of extruded boards is, of course, a continuous process but particle mixes can also be batch moulded to particular shapes; this process is often used for the formation of homogeneous mouldings for radio and television cases, and for moulded pallets and boxes.

Fibre board

Particles can, of course, be sub-divided further into individual fibres or bunches of fibres. This principle is used in papermaking but the same basic process is used in the formation of fibre boards. After chipping the wet particles are passed to defibrators and refiners, high shear machines that tear the particles apart. Alternatively the wet particles are passed to an autoclave where they are steam heated at a pressure of about 350 lbs./in.^{2} (2,300 kN/m.^{2}) for about 30 seconds and finally to about three times this pressure for a few seconds before the pressure is released. The water trapped within the individual chips then vapourises instantaneously into steam, tearing the particles apart to produce hydrolized pulp; exactly the same process is used for the production of puffed wheat! The pulp produced by either method is then spread on a continuous woven wire belt so that excess water can drain away; this water contains dissolved sugars as well as suspended starch and hemicellulose so that it is usually returned to the defibrators or refiners to ensure that these solids are not lost. In the production of low density insulation board the mat produced in this way is simply dried but in hardboard manufacture it is pressed to obtain the required density and surface finish.

Board properties

Plywood, particle board and fibre board all possess the normal cross-sectional movement of wood in the thickness of the board, with the single exception of the comparatively rare extruded particle board which has limited usefulness because of its low strength. In plywood the cross-sectional movement of the wood is restrained by the stable longitudinal dimension of the adjacent veneer so that plywood has very low movement in the plane of the panel, although it should be mentioned that excessive changes in moisture content can induce severe stresses in individual veneers so that they tear away from the glue line. In particle board some restraint occurs, particularly with long flat chips, as the random orientation ensures that the cross-sectional movement in some chips is restrained by the longitudinal dimension in the neighbouring chips but generally there is some movement in the plane of the panel, although not as much as in the cross-section of solid wood. Fibre boards are often somewhat more stable than particle boards but, as these products are essentially papers which rely upon natural association forces for their cohesiveness, there is a tendency for them to lose strength at higher moisture contents.

Processed boards

There is now an increasing tendency for all these panel materials to be manufactured with special finishes. Decorative veneers have, of course, been offered on plywood and chipboard for many years but it is now possible to obtain primed and painted panels, or special surface finishes such as phenol formaldehyde resin film faced plywood, originally developed for shuttering for casting concrete but now more widely used as a decorative or protective finish. Glass fibre reinforced plastic, rubber sheet and metal sheet finishes can be provided where durability or high wear resistance is required, and plywoods in particular can be supplied treated to preservative or fire-retardant specifications. In many senses these pre-finished products simplify assembly, although it will be explained in Chapter 4 that they are only as reliable in service as the joints between the individual panels; special tongued and grooved joints can be provided on plywood and chipboard panels to enable them to be efficiently used as load bearing flooring.

Compressed and resin impregnated wood

Finally, solid wood can be compressed or impregnated with resins in order to increase its density and to give greater stability. In theory these processes greatly increase the usefulness of wood, yet wood processed in this way is extremely expensive and the cost can seldom be justified. The value of resin impregnation treatments as flooring will be discussed in Chapter 4 but the principal limitation lies in the difficulties that are encountered in obtaining penetration into most species of wood. Clearly these processes are only realistic if particularly permeable species are employed. In Finland hardwearing birch flooring blocks have been produced by resin impregnation techniques and have been used, for example, very successfully for the floor of Helsinki Airport where there is a very high foot traffic load. In many species it is comparatively simple to penetrate the end grain and this principle has been used in Belgium in the manufacture of heavy duty industrial floors impregnated with pitch and with decorative floors in which hexagonal floor blocks are prepared from end grain slices of pine.

3. Wood Protection

3.1 Wood degradation

It must be accepted that wood is perishable. Indeed, if this were not the case our forests would be cluttered with the useless skeletons of dead trees. Unfortunately the various wood destroying insects and fungi are unable to distinguish between forest waste and wood in useful service. In the past wood was used principally in forest areas where it was readily available. Degrade was accepted but replacement was comparatively simple and inexpensive. In contrast wood is now a valuable commodity, transported considerable distances between the production forests and the ultimate user. It is essential for wood to be utilised efficiently in order to conserve world resources but also to avoid unnecessary cost, both to the individual user and to importing nations as a whole.

Declining natural durability

Wood has been an article of commerce for many centuries, some areas exporting exotic decorative woods and others supplying straight and strong structural woods. In the past high transportation costs were only justified for the most valuable woods, and even normal construction wood imported, for example, into Great Britain in the nineteenth century was generally slow grown and free from sapwood. Dwindling resources have since resulted in the introduction of thinning to encourage maximum yield, giving rapidly grown trees with wide rings, principally composed of springwood and, as the trees are felled when comparatively small in diameter, a large proportion of the wood is now sapwood; Swedish Redwood, Scots pine, is now approximately 50% sapwood. Whilst the strength properties are not significantly affected by wide rings and the presence of a high proportion of sapwood, the wood is far less durable and far less stable than the slow grown heartwood that was commonly used in the past.

Biodegradation

Whilst fire and moisture content changes must be clearly recognised as causes of degrade in wood it is, of course, the various living organisms, both microbiological and animal, which are perhaps most important. The living tree in the forest passes through a series of stages, originating with the death of the living tissue and then passing through a drying stage until the protective bark is lost so that rewetting is then able to occur. Commercial wood passes through similar stages and it is possible to establish a sequence of degrading organisms to which wood is susceptible at the various stages in its progress from the living tree to the wood in service and then beyond to the various stages of neglect. It is impossible in a general book on "Wood in Construction" to describe even the important wood destroying organisms in detail and this chapter must therefore be confined to explaining their significance. Literature describing local wood destroying organisms, particularly insects and fungi, is available from government departments and appropriate associations in most countries, and this subject is covered in detail in "Wood Preservation", a companion volume to this book, written by the same author.

Wet green wood

Freshly felled green wood has a very high moisture content, 60–200% for most species but as high as 400% in extreme cases. Sugars and starch are present, particularly in the phloem and parenchyma tissue respectively, and these conditions are particularly attractive to the sapstain fungi, such as *Ceratostomella* spp. The hyphae, the minute strands comprising the growing tissue of a fungus, are brownish in colour but when affected by a mass of these hyphae defraction gives wood a blue or black colouration, often known as blue staining or blueing. Sapstain problems can be largely avoided in coniferous woods by winter felling which ensures a lower moisture content and virtual freedom from sugars. However floating can be the cause of staining problems and, in wet wood, an increase in temperature during shipment can cause sweating if the cells are still alive and the oozing of sugar-rich sap from the ends of the log and the development of a luxuriant fungal growth. In fact, sapstain fungus is comparatively unimportant as it does not destroy the wood cell or cause any structural damage but it is unsightly and leads to downgrading and thus reduced value.

Sapstain

Stain control

In recent years the excessive demand for softwood has caused the felling period in the coniferous forests to be extended throughout the whole year instead of being confined to the winter months alone. Wood felled during the active growing season, particularly the early spring, contains both higher moisture and sugar contents than winter felled wood and there is a corresponding increase in the danger of staining. Indeed staining is virtually inevitable for fellings in late spring or early summer, unless they are converted and kiln dried immediately. Even then there is still a danger that stain will develop during the early stages of kilning before the moisture content has been significantly reduced. The best solution to the problem is to convert the logs immediately into sawn timber which is then given a stain control treatment. Alternatively the sawn wood is transferred immediately to kilns for drying and a toxic treatment is introduced into kiln atmosphere. The most widely used stain control treatment is sodium pentachlorophenate which is significantly volatile, particularly when treated wood is drying, so that much of the treatment is certain to be lost during the kiln drying process. A treatment is therefore desirable when drying is complete in order to prevent the recurrence of staining should the wood become accidentally rewetted. In many cases this final treatment is not used but instead the wood is packaged and wrapped in plastic sheeting in order to retain the low moisture content achieved by kiln drying and thus to prevent the development of staining. In fact, packaged wood wrapped in this way can often become very seriously affected by staining, particularly the outer sticks in contact with the plastic, should it be exposed to temperature changes which can induce evaporation of moisture from the wood during warm conditions but condensation under the plastic wrapping during a subsequent cold period. For this reason packages are frequently dipped in a stain control treatment before wrapping but this has only limited success as it can give little protection to the inner pieces within the package. The only proper solution is to treat pieces individually before packaging and wrapping, yet the commonly used sodium

Stain in wrapped dry wood

pentachlorophenate treatment is aqueous based and treatment will necessarily increase the moisture content of the wood, partly defeating the object of the original kiln drying. Clearly there is scope for the development of improved and more reliable stain control treatments.

Bark borers, wasps and longhorn beetles

Bark borers frequently attack freshly felled logs. The wood wasps, particularly *Sirex noctilio*, have attracted particular attention in recent years, largely through the fears of the Australian authorities that wasps introduced in softwood packaging from the northern hemisphere would damage valuable pine plantations; all wood imported into Australia is subject to quarantine regulations and must either be inspected or destroyed on arrival, or must be accompanied by a certificate to show that it has received an approved treatment. Wood wasps bore through the bark with their ovipositer and lay their eggs in contact with the phloem. The larvae hatch from the eggs and then explore the phloem, living on the sugar content in the sap and, at the same time, loosening the bark; this is the first stage in the destruction of a dead tree in the forest. Some Longhorn beetles behave in a very similar manner, laying their eggs in cracks in the bark. The developing larvae either explore the phloem in search of nourishment in the same way as the wood wasps, or alternatively they tunnel in the sapwood, gaining nourishment from starch deposits or, in a few cases, the cellulose of the wood.

Ambrosia beetles

Pinhole or shothole damage

Both the wood wasps and Longhorn beetles can be quite large, with a full grown larval size averaging perhaps 1″ (2.5 cm.) or more in length. In contrast the Ambrosia beetles are much smaller, only about $\frac{3}{8}$″ (3 mm.) long. These insects, the Scotylidae and the Platypodidae, tunnel through the bark to produce extensive galleries in which they lay their eggs. At the same time they introduce fungi which thrive in the high moisture, sugar and starch content of the freshly felled wood, thus providing the larvae with food; the browsing of the larvae on the fungus accounts for the name Ambrosia. The galleries either extend under the bark or into the sapwood, perhaps following the less dense spring wood of the annual rings, and the distinctive pattern of the galleries is frequently a characteristic of a particular species. Ambrosia attack can be very severe in the tropics, frequently introducing stain as the Ambrosia fungus which, even if it does not extend to the entire sapwood, results in a characteristic streak of stain along the grain on either side of each gallery. The damage to the wood is often described as pinholes or shotholes, depending upon size, and the damaged wood is known as pinwormy. Because of the danger of Ambrosia beetle attack coupled with staining it is normal to spray logs immediately after felling in the tropics with a mixture of a contact insecticide and a stain control treatment, whilst sawn wood produced close to the forest before shipment also receives a further stain control treatment as it is normally shipped with a relatively high moisture content.

Fresh dry wood

Powder Post beetles

As the moisture content of the wood falls it becomes immune to fungal decay but it may be susceptible to Powder Post beetle damage. The Bostrychidae adult beetles bore tunnels into the sapwood in which they lay their eggs, the hatching larvae relying upon stored starch in the wood tissue for their nourishment. The Lyctidae in contrast lay their eggs in large vessels or pores, although again the hatching larvae depend upon the starch content in the wood. Powder Post attack is most common in hardwoods, although Bostrychid attack can occur in both softwoods and hardwoods whereas Lyctid attack is confined to large pored hardwoods. Powder Post beetles produce flight holes about 1/16 in. (1.5 mm.) diameter. All wood is immune to Powder Post beetle attack if it is free from starch. This means that, for

Figure 3.1 Lyctus Beetle, *Lyctus brunneus*

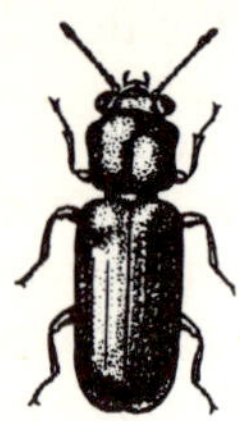

example, temperate hardwoods are most susceptible if felled in the early winter when the starch content is at its highest level. Starch appears to naturally degenerate slowly so that wood is virtually immune several years after felling. In addition air seasoned wood is far less susceptible than kiln dried, apparently because the cells remain alive during air seasoning and thus utilise the starch whereas the cells are killed in kiln drying so that the starch deposits remain and attract Powder Post beetle attack.

The Anobiidae, the Furniture beetles, are a very important family. The bark borer, *Ernobius mollis,* lays eggs in cracks in the bark of dry softwood logs, or bark still attached to the waney edges of sawn wood. The hatching larvae explore the phloem in search of nourishment, loosening the bark, but the galleries sometimes extend for a short distance of up to $\frac{1}{2}''$ (1 cm.) into the sapwood. These holes in the sapwood are sometimes confused with those of the Common Furniture beetle, *Anobium punctatum,* but they tend to be rather larger in diameter and they are invariably associated with galleries under bark; the bark may have fallen away from the adjacent waney edge but the galleries sometimes extend for a short distance of up to $\frac{1}{2}''$ (1 cm.) into the beetle is the most important aspect of *Ernobius mollis* attack. It is dependent upon sugar in the phloem and, to a lesser extent, starch in the sapwood which deteriorates naturally a year or two after conversion so that *Ernobius* is eliminated naturally without any need for treatment.

Dry wood

If small holes are observed in the sapwood of a piece of softwood with no evidence of either bark or a waney edge, then the Common Furniture beetle, *Anobium punctatum,* is likely to be responsible. In fact, Common Furniture beetle attack will occur in the sapwood of most softwoods and hardwoods, as well as the heartwood of some temperate hardwood species such as birch and beech. Eggs are laid, usually in the summer months, in cracks on the surface of the wood or, in the case of furniture, perhaps in open joints. The larvae hatch out by breaking through the base of the eggshell and boring straight into the adjacent wood. The insect usually remains in the larval stage for almost a year before forming a chamber just below the surface in which the full grown larva, about 1/5″ (5 mm.) in length, pupates into an adult beetle. The adult bores to the surface and escapes by a circular flight hole about 1/16″ (1.5 mm.) in diameter. The adults mate and then the female induces a fresh attack on the wood by laying eggs, sometimes in the old flight holes. Obviously the damage is caused entirely by the larvae in boring through the wood, in contrast to the Ambrosia beetles where the adult beetles are responsible for tunnelling.

Figure 3.2 Common Furniture beetle, *Anobium punctatum*

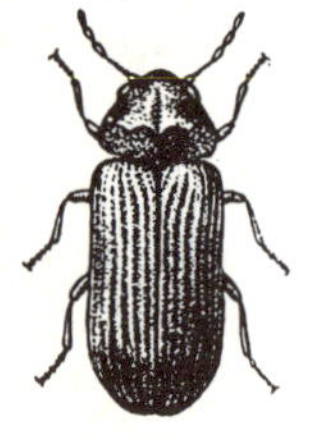

Damp wood

If wood becomes damp it will considerably encourage the activity of the Common Furniture beetle, enabling it to attack heartwood which would otherwise be resistant. It appears that, although fungal attack may not be obvious, fungal or bacterial activity is converting the wood to a form that is more readily assimilated by the insect. The Death Watch beetle, *Xestobium rufovillosum,* is entirely dependent upon the presence of microbiological attack. Indeed wood attacked by the Death Watch beetle is often brownish in colour, indicating perhaps prolonged incipient fungal attack although there may be no other evidence. The Death Watch beetle is far larger than the Furniture beetle and, in the British Isles for example, it is generally associated with historic churches, apparently because the periodic heating combines with the traditional lead roof covering to encourage condensation, incipient fungal attack and thus Death Watch beetle attack in the roof sarking boards. However, the rather sinister name of the Death Watch beetle is not

Figure 3.3 Death Watch beetle, *Xestobium rufovillosum*

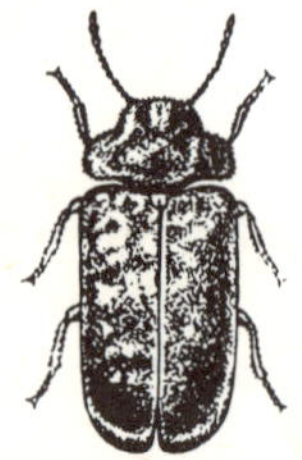

associated with churches in any way but with the characteristic tapping which is adopted by the adult beetles as a mating call and which is produced by striking the top of the head against the surface of the wood on which the beetle is standing. On a calm day the sound is distinctly audible in a quiet building, perhaps particularly one that is silent through the presence of death.

Whilst it is logical to treat the Anobiidae as a group progressing from the bark borer to the Furniture beetle attacking dry wood and ultimately the Death Watch beetle dependent upon some decay, this system unfortunately displaces the House Longhorn beetle from its rightful position in association with the Common Furniture beetle in this sequence of attack. The House Longhorn beetle, *Hylotrupes bajulus,* is an extremely important wood borer, only associated with the Common Furniture beetle in this description in the sense that it attacks similarly dry wood in buildings. The attack is confined to the sapwood of dry softwoods and is thus of particular importance where wood with a large sapwood content is used for structural purposes. The life cycle, basically similar to that of the Common Furniture beetle and, of course, all other wood boring beetles, varies from three to eleven years. This is a large insect with a fully grown larva perhaps $1\frac{1}{4}''$ (3 cm.) long. A female beetle may lay up to 200 eggs and, if most of these hatch within the same roof or floor structure, very substantial damage can be caused during the years before the adult beetles ultimately emerge. Indeed, the entire sapwood may be riddled by oval galleries, leaving a thin intact veneer over the surface of the piece of wood. The adult beetle emerges from the wood through an oval exit hole about $\frac{3}{8}''$ (1 cm.) across, but the presence of this flight hole is a certain indication that severe damage has already been caused and the structural integrity of the piece of wood must be checked by probing.

Figure 3.4 House Longhorn beetle, *Hylotrupes bajulus*

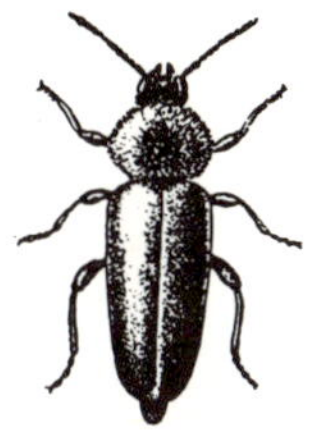

The Death Watch beetle is only one of several wood borers which are dependent upon damp or wet conditions in order to attack wood. The presence of wood boring weevils is often indicated by the presence of small holes in damp wood attacked by Death Watch beetle or Common Furniture beetle. If large pieces of relatively durable wood, such as European oak, remain in a damp and slowly decaying condition for a prolonged period they may attract the attention of other insects such as *Helops coeruleus,* a large blue beetle with long larvae possessing two recurved spines on its tail, or even Stag beetles which are more commonly found in decayed roots or fence posts below ground level. These insects are, of course, dependent upon fungal decay which is a far more significant factor in wood degradation than the insect damage. Fungal damage can, in fact, originate within the standing tree. Whilst the very high moisture content in the living sapwood tissue generally prevents fungal decay and insect attack, physical damage to the bark and cambium or scars from the removal of branches may permit the moisture content to fall to a level where fungal or insect attack is possible. Branch and root scars are the particular danger points as they may expose relatively dry heartwood which, whilst it is relatively durable in most wood species, will permit decay in others. In the case of a branch scar the first stage is probably rapid drying from the exposed end grain and the development of checks. Subsequent rainfall is then trapped within these checks and produces a gradient of moisture content which permits spores in the atmosphere to seek out precisely the conditions they require for germination. The fungal infection then spreads progressively through the heart of the tree. These heart rots are variously known as dote and punk, or they are referred to by the name of the attacking fungus, such as *Armillaria mellea,* the Honey fungus.

Figure 3.5 Wood weevil, *Pentarthrum ruttonii*

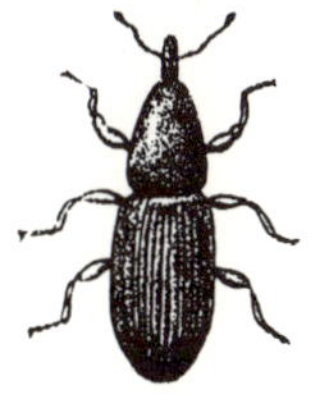

Fungal decay in trees or logs

There are no obvious external signs of damage until the decay is far advanced and a sporophore, or fruiting body, appears through the bark, usually indicating that the heartwood of the tree is virtually hollow. In dotey heart the attack is widespread but damage is limited, perhaps because the heartwood moisture content was too low. Whilst the infection will be killed if the wood is kiln dried at a sufficiently high temperature, failure to take this precaution introduces the danger that the infection will become reactivated if the moisture content reaches an adequate level. For example, logs are frequently stacked in the open and, if they are dotey, they are likely to become covered with the fluffy white hyphae or growing elements of the reactivated fungus. Washing floors can also reactivate dote, the upper surfaces softening progressively until they eventually break up and fail. Dote normally causes a colour change in the wood, most frequently to a brown colouration in the centre of the heart, although occasionally off centre when the effect is referred to as a false heart. There is a smell characteristic of the fungus involved.

The only special feature about dotes is their origin in the standing tree or in a log that has been permitted to remain for a protracted period within the forest. In all other respects dotes are characteristic of the most important group of wood-destroying fungi, the Basidiomycetes. In the case of brown rots the fungus destroys the cellulose, leaving the lignin which gives the wood a characteristic brown colouration and usually cross grain cracking. Common dote, *Trametes serialis,* is a brown rot but other dotes are described as white rots because the fungus decays both the cellulose and lignin, generally resulting in less colour change and a fibrous appearance. *Fomes annosus* is a dote found in spruce, larch and red oak heartwood, giving a typical reddish or purplish colouration. The purplish colour develops particularly in spruce but the colour is lighter in larch. The attack is typical of dote, appearing on the cut face as tiny white pockets of rot filled with growth like small pieces of cotton wool which become more fibrous as the decay progresses. If dote is suspected but not apparent in this way it can usually be detected by lifting the fibres with a knife; if they are long and springy the wood is sound but if they are brash or break readily across the grain then decay must be suspected.

Brown and white rots

Fungal decay in wood in service

Any dote activity that redevelops in wood in service can be ultimately attributed to the germination of a spore, perhaps on a branch scar on a standing tree or a log lying in the forest. The other principal decays of wood in service differ only in the fact that spore germination occurs on the surface of the wood when the surface conditions induce an appropriate moisture content within the wood and relative humidity in the surrounding atmosphere. A vast selection of spores are invariably present in the atmosphere and, if suitable conditions occur, an appropriate fungal infection will surely develop. For example, a post standing in moist ground will generally provide all conditions varying between the high moisture content in the ground and the low moisture content in the well ventilated aerial parts of the post, although it is the intermediate conditions at the ground line that attract the most serious decay; in the forest this ensures that the relatively dry tree falls to the ground where undergrowth inhibits evaporation and the increasing moisture content accelerates the process of decay.

Design precautions in buildings

There are a very large number of wood destroying fungi encountered in buildings throughout the world. However, it must be emphasised at this point that there is never any need for decay in building timber as simple design precautions are usually adequate to ensure that the wood never becomes sufficiently wet to support decay fungi. There are some special conditions, such as external window frames or piles in the ground, where decay is highly probable but it can be avoided by taking proper precautions such as the selection of suitable durable wood or the use of an adequate preservative treatment. Despite the ease with which it can be avoided, fungal decay frequently occurs in buildings through neglect in design or maintenance. The cellar rot, *Coniophora cerebella (puteana),* is a particularly good example of a decay that can occur as a result of neglect in design or construction. It occurs in persistently damp conditions, as when a damp proof course is omitted and when plates under floor joists are in direct contact with damp supporting walls. If the moisture content tends to fluctuate, as in wood affected by a periodic roof leak, the white pore fungus, *Poria vaillantii,* is far more common in the softwoods and, for example, the stringy oak rot, *Phellinus megaloporus,* in oak. *Polystictus versicolor* sometimes develops when non-durable tropical hardwoods are used as drips or sills on external joinery (millwork), and *Paxillus panuoides* generally occurs where the conditions are too wet for the cellar fungus.

Wet rots, in persistently damp conditions

These are only a few of the fungi that may be encountered but they serve to illustrate the way in which a particular fungus is perhaps associated with both a particular type of wood as well as particular conditions. The *Coniophora, Poria* and *Paxillus* spp. are all brown rots, giving the affected wood a distinct brown colouration and a varying degree of cross grain cracking. The *Phellinus* and *Polystictus* spp. are white rots, causing only limited change in the colour of the wood but very pronounced softening and loss of strength.

Dry rot, *Merulius lacrymans*

Perhaps the fungus that is most widely known as causing serious damage in buildings is the dry rot fungus, *Merulius (Serpula) lacrymans.* Dry rot spores will only germinate when the atmospheric relative humidity is suitable, generally where accidental wetting, perhaps from a plumbing leak or a roof defect has allowed the wood to become very wet. Subsequent drying, perhaps a seasonal effect or through correction of the defect, may permit the relative humidity to reach the optimum level so that spore germination can occur. Provided the ventilation is limited the relative humidity will fall only slowly and this will permit the fungal infection to become established. Spore germination consists of the development of hyphae, or threads, which penetrate into the wood, radiating from the original point of germination and branching so that the affected area is covered with a soft white growth like cotton wool. If the conditions are favourable successive masses of hyphae will become compacted on the surface of the wood, particularly if there is occasional drying, to form a dense skin or mycelium.

The growth will not be confined to wood but will spread across and through plaster, brickwork, stone masonry and concrete in an attempt to discover further supplies of wood for nourishment. Obviously this exploratory growth must be provided with food from adjacent wood attacked by the fungus and the hyphae develop into rhizomorphs, or conducting strands, which both convey food from wood to the exploring hyphae and perhaps water obtained from damp masonry. The fungus will use water absorbed in this way and formed during the assimilation of cellulose to form globules on the surface of the growth which, if the ventilation is restricted, will maintain the humidity at the optimum level desirable for growth. This habit of forming "tears" on the surface of the growth explains the botanical name *lacrymans* and the description in the Old Testament Book of Leviticus,

Chapter XIV, of dry rot as the "fretting leprosy of the house". As the attack progresses the cellulose is destroyed, giving the wood the characteristic dark brown colour of a brown rot, accompanied in the case of dry rot by very pronounced cross cracking which, in combination with longitudinal cracking, gives the wood a characteristic cuboidal appearance. The preference of dry rot for unventilated situations in which it can control the humidity ensures that it generally remains largely concealed and the first sign of damage, other than a characteristic odour, is perhaps the buckling and the cracking of a painted skirting. By then the attack may well be very extensive, perhaps spreading through masonry for considerable distances in all directions in its search for wood. When the wood supply is nearing exhaustion or when the humidity falls unexpectedly the fungus may spread onto the surface of the concealing wood or plaster and form a sporophore, or fruiting body, which will produce millions of red-brown spores in an attempt to infect any other wood that may be in a suitable condition in the vicinity.

Wet wood

Soft rot

Wood that is continuously immersed in water becomes saturated and immune to the attack of Basidiomycetes, the brown and white rots, but soft rot can occur, caused by Ascomycetes and Fungi imperfecti. Soft rot takes the form of a softened layer of wood on all the exposed surfaces, the damage progressively increasing in depth at a very slow rate. Unprotected wood immersed in fresh or sea water is invariably affected in this way but the damage is comparatively insignificant in large section timbers such as those used for piling and the construction of groynes. In wood of thinner sections the loss in strength can be significant, as in a neglected boat where planking has been exposed through abrasion damage to the paintwork or in cooling tower slats where the high temperature results in particularly rapid soft rot attack in inadequately preserved wood.

Marine borers

Continuous immersion in sea water also introduces the danger of marine borer attack. Several animals can infest wood in this way but damage is most commonly due to Crustaceans called gribble, *Limnoria,* and Molluscs called shipworm, *Teredo*. Gribble attack takes the form of superficial burrowing into the wood. Small holes less than 1/10″ (2.5 mm.) diameter are produced and extensive attacks seriously weaken the surface of the wood which in time becomes worn away, exposing fresh surfaces to attack. This is a major marine pest in most parts of the world on account of its widespread distribution. Shipworm attack is very difficult to detect. The animal enters the wood by boring a hole about 1/50″ (0.5 mm.) diameter. This is then extended to form a long tunnel with a characteristic calcareous lining. Severe attack may considerably reduce the strength of the wood but this may not become apparent unless abrasion or gribble attack exposes the burrows and their characteristic linings. Shipworm is generally confined to relatively warm and saline waters so that, in Europe for example, it generally occurs on coasts where the water is warmed by the gulf stream and it is absent in river estuaries and the Baltic Sea.

In boats most normal antifouling paints provide efficient protection against marine borer attack but, in areas where marine borers are a particular problem, regular inspections are advisable to ensure that unprotected wood is not exposed by abrasion. For example, wooden rudder trunkings cannot be effectively coated with antifouling paint and shipworm frequently becomes established as a result. Paint cannot be applied to heavy marine woodwork and naturally durable or adequately preserved wood must be used. In fact, it is relatively easy to preserve against shipworm attack but gribble is a far greater problem; most preservatives will reduce the rate of attack but it is not always entirely prevented.

Anaerobic waterlogging

If wood is waterlogged and completely surrounded by, for example, an impervious clay or mud layer there is no possibility of fungal decay or borer attack because of the complete lack of oxygen. However, some anaerobic bacteria are able to survive in these conditions, obtaining the necessary oxygen by their reducing action on suitable chemicals that may be present, such as by the conversion of sulphate to sulphide and thus the generation of the hydrogen sulphide smell which is a typical feature of these conditions. These bacteria do not cause any damage to wood, although generally it becomes very dark brown in colour. However, protracted waterlogging results in slow hydrolysis of the cellulose which is progressively lost. The amount of loss could be gauged by assessing the dry density of the wood but this is very difficult as any attempt to dry the sample invariably results in considerable distortion, often rather reminiscent of the shrinkage that occurs from brown rot attack which is, of course, another method for removing the cellulose from wood and leaving the brown lignin. The alternative is to measure the saturated moisture content of the sample by weighing it firstly when wet and subsequently when oven dried. The amount of water loss is then related to the dry mass to give the saturated moisture content and this can be directly related to the period of immersion. This technique is sometimes used by archaeologists to estimate the age of an ancient structure when they are able to obtain waterlogged wood samples, for example from piles driven into a bed of clay.

Termites

Termites, or white ants as they are commonly called, are probably the most serious wood destroying pests. They are not ants but belong to the Isoptera, whereas true ants are Hymenoptera, an order which also includes the bees and wasps. However, like the true ants, the termites are social insects living in communities with specialised forms or "castes", the workers and soldiers, as well as male and female reproductive individuals. Termites are widespread in tropical and sub-tropical countries, and the rate and severity of their attack makes them a serious pest and economically significant wherever they occur. There are approximately 1,900 identified species of termites, 151 known to damage wood in buildings and other structures. With such a vast number of species involved it is essential when considering preservative systems to have a knowledge of the basic behaviour and differences between the various families.

Termites are principally tropical in distribution but are encountered as far south as Australia and New Zealand and as far north as France. Improvements in transportation and world trade have been responsible for the wider distribution of termites and this is particularly apparent in France and Germany. For example, the Termite of Saintonge, *Reticulitermes santonensis,* is established on the west coast of France between the rivers Garonne and Loire but it has spread to Paris where it is concentrated around the Austerlitz station which serves this coastal region. In a similar way *Reticulitermes flavipes* occurs in Hamburg, although the species are both sensitive to climatic conditions and they are not expected to spread widely; they are largely concentrated in heating pipe ducts and other rather warm areas in buildings. However, it is still remarkable that the British Isles remain completely free from termites.

A common feature of the six families of termites within the order Isoptera is the lack of cellulose in their digestive enzymes, despite the fact that all six families possess members which are wood destroying. Three of the families are of only limited significance; the Mastotermitidae are represented by only a single primitive species in Northern Australia, the Termopsidae include three species that infest buildings in the United States of America, and the Hodotermitidae are confined to semi-desert areas of South Africa, North Africa and the Middle East. The remaining three families are of considerable significance as they contain the more important wood destroying termites.

The family Termitidae is a large and mixed group which includes the subterranean and the mound termites which construct nests under the ground, on the sides of trees or as mounds on the ground. They are able to digest wood without the assistance of an intestinal symbiont because they rely upon fungus for the production of a cellulose to convert the wood to a form suitable for their own digestion. The Termitidae can be divided into two distinct groups. The first group contains the Microcerotermes, Amitermes, Nasutitermes etc., which depend upon prior infection by a fungus to convert cellulose to a digestible form. Lignin is generally unaffected and excreted, providing the raw material for the construction of the typical honeycomb nests and covered walkways. In contrast the group containing the Macrotermes, Odontotermes, Microtermes, etc., is not restricted to wood already infected by fungus but instead these convert all cellulose to a digestible form in "fungal gardens" in their nests. These termites gnaw fresh wood into fragments which are then chewed to paste and excreted in the nest where they become infected by the termite with a fungus which causes deterioration of both cellulose and lignin. The wood is therefore converted to a form which is readily digested by the termites.

The family Kalotermitidae includes the dry wood termites which, as this name implies, are able to destroy wood possessing a very low moisture content. A symbiont Protozoa in the hind gut provides cellulases sufficient to enable these termites to digest wood cellulose in a normal manner, although lignin is unaffected and is excreted. Colonies of these termites inhabit sound dry wood and rarely enter the ground; for this reason all the other families are sometimes collectively known as subterranean termites. The attack is spread by winged egg-laying females and, in areas where there is a risk of dry wood termite attack, it is essential to use naturally durable or adequately preserved wood.

The family Rhinotermitidae are sometimes described as the moist wood termites; in contrast the minor family Termopsidae are sometimes described as damp wood termites. The Rhinotermitidae cause damage to buildings and other structures but are of far less economic significance than either the Termitidae or the Kalotermitidae. They possess protozoan intestinal symbionts, but they prefer moist wood which is already infected by fungi or bacteria, thus achieving more effective digestion and assimilation than in the case of the dry wood termites.

In many areas the main termite hazard is confined to the subterranean species of the Termitidae. The first group of this family depends upon prior infection of the wood by fungus. They are therefore a particularly serious hazard to fence posts, transmission poles and all other wood in soil contact. However, attack can be readily prevented by taking the conventional precautions to avoid fungal decay; ensuring that the wood remains dry in buildings and using only wood that is naturally durable or adequately preserved in ground contact conditions. In the case of the second group control is more difficult as they will physically destroy undecayed dry wood, ingesting fragments and transporting them back to the nest where they are excreted, processed in the fungal gardens, and thus converted into useful food. Physical barrier systems provide the most common method for protecting building structures from this particular hazard. The Termitidae are unable to fly and protection can be obtained by isolating wood from the soil by the use of shields of metal or plastic between the wood and the footings, by introducing a barrier of poisoned soil or concrete, or by painting structures white if they provide a route to the wood; the termites dislike constructing their walkways on light coloured surfaces. These barrier systems are reasonably effective provided they are conscientiously constructed to prevent the termite discovering or constructing an alternative route to the wood structure. In particular, it must be appreciated that subterranean termites are capable of constructing tubular walkways spanning distances of 1′ (30 cm.) or more and this may enable them to bridge termite shields. In addition barriers of this type are completely useless if bridged by negligence in subsequent construction or maintenance, such as the careless installation of electric cables and plumbing.

Many other termites, particularly the dry wood termites, the Kalotermitidae, are able to fly and cannot be realistically controlled by physical barrier systems. If wood is to be used in structures exposed to attack it must be naturally resistant or adequately treated with preservative if it is to survive.

3.2 Wood preservation

It is possible to find woods with natural resistance to almost all destructive organisms, the heartwood generally being more durable than the sapwood. However, the physical properties, availability and cost may preclude selection in this way. In buildings in particular structural precautions to prevent wood becoming wet will be sufficient to avoid the major hazards of fungal attack and associated insect infestation. The use of damp proof courses to isolate floor joists from soil dampness, overhanging eaves and efficient rainwater disposal systems are examples of structural precautions significant in decay prevention. It is often claimed that painting wood, particularly external joinery (millwork), protects it from rainfall and thus preserves it against decay. In fact, minor imperfections or damage to the paint film will permit absorption of water whilst the remaining paint "protection" will simply restrict evaporation and thus causes the dampness to accumulate. Joinery frequently has only a single primer coat as protection on the hidden faces, despite the fact that these are precisely the areas where most damage occurs during installation and where contact with adjacent damp materials may result in moisture absorption.

Structural precautions

In fact, the most important preservative action of paint is as a barrier, isolating the wood from attacking insects or fungi. Minor damage may permit access to be obtained to the wood and protection is clearly more efficient if the wood is impregnated in depth rather than simply coated. This is the origin of pressure treatment with coal tar distillates such as creosote, but all treatments of this type cause a fundamental change in the appearance of the wood; to the attacking organism the wood has the appearance of being a solid block of creosote or paint, and the wood has this same appearance to the human observer. Such changes may be aesthetically unacceptable and, if the treatment is considered dirty, unsuitable for use in many situations. The alternative is a complete abandonment of the barrier principle in favour of a toxic action, low retentions of highly toxic compounds achieving preservation without pronounced alterations in the physical properties and appearance of the treated wood. This principle was first applied in the form of the salt preservatives such as zinc chloride, mercuric chloride and copper sulphate. These simple salt treatments possess very poor resistance to leaching and, although suitable salts may give adequate protection against insect attack in dry conditions, they are generally quite unsuitable for use as fungicidal preservatives. A variety of multi-salt preservatives have now been developed with the intention of improving fixation, usually by the addition of chromates, or extending the spectrum of activity, for example by the addition of arsenic to improve preservative action against insects. These aqueous multi-salt preservatives are giving excellent service throughout the world but they suffer from two distinct disadvantages. They must be applied to wood at low moisture content in order to achieve the necessary absorptions of preservative solution, yet they result in treated wood of exceptionally high moisture content which must be reduced before it can be used for many purposes. These changes in moisture content can result in severe distortion and possibly a high rejection rate. The cost of solvent alone is sufficient to ensure that organic solvent preparations are more expensive, yet they are progressively replacing the aqueous multi-salt preservatives wherever these factors are of importance, as in the treatment of joinery (millwork).

The selection of an appropriate preservation process will, of course, depend upon the situation in which it is to be used. Vacuum and pressure impregnation treatments may give improved penetration which will be favoured whenever there is a danger that a shallow treated zone may be damaged. Examples are woods, such as marine fenders, which are subject to physical damage and others which are subject to severe insect "tasting" damage, the erosion that occurs, particularly with termites and gribble, when a certain amount of treated wood is removed before a lethal dose is acquired and causes death. These are arguments against the use of "superficial" treatments applied by brush, spray or immersion. However, many woods are distinctly resistant to impregnation, even at high pressure and the superficial treatment based on a penetrating organic solvent may be just as effective, even on woods that are considered perfectly suitable for pressure impregnation, such as Scots pine, *Pinus sylvestris,* in which penetration rarely occurs into the heartwood. Re-sawing, cross-cutting, jointing and drilling will expose untreated zones in woods that possess heartwoods that are resistant to impregnation and many manufacturers therefore recommend brush application of concentrated preservative solution to these exposed areas. These are generally end-grain areas which, through their porosity and tendency to absorb water, are particularly susceptible to decay. As a result the most critical part of a structure is often protected only by a brush treatment casually applied on site. It is sometimes argued that more reliable preservation can probably be achieved by dipping of wood on site after cutting to size. This is perhaps true but the same facts can also be interpreted in favour of the argument that there is no point in pressure impregnation unless it is applied to wood after cutting to size and after all working has been completed.

It is impossible in a book of this type to cover the subject of wood

Forest and mill treatment

preservation in detail; instead this is the subject of a separate companion volume, "Wood Preservation". It is apparent from the descriptions in the previous section of this chapter that there are three principal degradation problems. The first is concerned with freshly felled wood possessing a high moisture content, sugars in the phloem and starch in the parenchyma tissue which makes it particularly susceptible to sapstain fungi, bark borers and, after drying, the Powder Post beetles. Sapstain is best avoided by rapidly kiln drying the wood but this kills the cells and the wood still retains both sugars and starch deposits, particularly in some hardwoods which thus become exceptionally susceptible to Powder Post beetle attack. In addition, stain will recur should wood subsequently become rewetted. The only reliable solution is the treatment of wood immediately after conversion, as well as the re-treatment of the wood after kiln drying if the kiln temperature is sufficient to volatilise the preservative compounds. In the northern coniferous forests of North America, Scandinavia and Russia stain is a very severe problem which leads to a reduction in value so that precautions are essential. Usually an aqueous solution of sodium pentachlorophenate is applied by a dipping process, although there are currently criticisms of the toxicity of this compound and there are attempts to develop alternative safer treatments. In the tropics the pinhole borers are particularly important, both because of the damage caused by their galleries and by the staining fungi that they introduce, and logs are often sprayed with a combined fungicidal and insecticidal solution immediately after felling to give protection. Again sodium pentachlorophenate is widely used, often in combination with an emulsion of a contact insecticide such as Lindane or Dieldrin. Finally, dry wood with a high starch content in the sapwood is likely to attract Powder Post beetles and, if it is desired to store the wood with the bark and sapwood attached, a contact insecticide emulsion treatment is essential; an alternative precaution is to ensure that all bark and sapwood is cut away and burned.

Construction wood treatments

The second basic preservation problem concerns structural members in a properly designed building in which the wood remains dry. Provided the building is properly maintained there is no danger of fungal attack but Furniture beetle, House Longhorn beetle and dry wood termites may still cause serious damage. The only danger in cooler temperate areas is Common Furniture beetle attack and, as this will be confined to the sapwood of structural softwoods, treatment cannot be justified if the members are predominantly heartwood. However, floor boards sometimes have sapwood edges which, if they are attacked by Common Furniture beetle, can collapse under the extraordinary compression loads that can be imposed by stiletto heels whenever they are in fashion, and some precautionary treatment may be considered desirable. House Longhorn beetle is a much more serious problem as it can rapidly cause complete destruction of the sapwood of softwood structural members so that, in areas where there is a danger of House Longhorn beetle attack and where there is a comparatively high percentage of sapwood in the construction, precautionary treatment is essential. For example, the House Longhorn beetle is confined to certain localities in England and South Africa, and regulations now require all structural woodwork to be adequately treated. In the Building Regulations for England and Wales the wood is required to be vacuum impregnated, or dip treated for not less than ten minutes, in an organic solvent wood preservative containing 0.5% by weight Lindane or Dieldrin. This treatment will give protection only against these insects but, in practice, proprietary preservatives usually contain fungicides to give protection against decay resulting from accidental leaks in buildings. Indeed the regulations also specifically permit borate diffusion and copper-chrome-arsenic impregnation treatments which are both insecticidal and fungicidal. The South African regulations cover only similar general purpose preservation treatments, although they are actually specific to the control of insect attack by Longhorn beetles and a dry wood termite, *Cryptotermes brevis*.

Treatments for wood in ground contact

Finally, the third area in which preservation is required is where wood becomes rewetted in service. The most important precaution is to avoid wetting by incorporating damp proof courses, overhangs and rain water disposal systems within the structure. There remains a danger of accidental wetting through, for example, roof and plumbing leaks, but these really represent an insignificant danger where buildings are properly maintained. However, there are situations in which wood, if not adequately protected or selected, will certainly decay, such as fence posts, transmission poles and piles where there is a severe risk of fungal decay at the ground line. In many climates wood used in marine construction will be subject to severe borer attack and in the tropical and sub-tropical areas there is almost always a danger of termite attack. In these situations the use of naturally durable or adequately preserved wood is widely accepted, and if this precaution is neglected, frequent and perhaps expensive replacement will certainly be necesssary.

Degradation classification

Preservative treatments can be classified in several different ways. One simple form of classification is related to the danger of degradation. Thus sapstain in freshly felled green wood, Powder Post beetle in fresh dry wood and Common Furniture beetle in structural softwoods are considered to represent only a slight hazard as they have very limited structural significance. However, where severe decay may occur as a result of accident or negligence the risk is considered to represent a moderate hazard; if the decay occurs it may be structurally serious, yet decay is not certain. An excellent example of this situation is external joinery (millwork) such as window frames. In the British Isles window frames are generally constructed from Baltic redwood, *Pinus sylvestris,* which has non-durable sapwood. The principle moulded sections, such as drips and sills, are often prepared from non-durable tropical hardwood such as obeche. Cracking of the paint film often occurs at joints, particularly where end-grain with its high cross-section movement is in contact with side-grain with its very low movement, and water penetration then occurs into the exposed joint, followed particularly by absorption into the adjacent permeable end-grain. The paint remaining around the crack ensures limited ventilation and thus the wood remains damp, so that the use of non-durable woods invariably leads to decay. Typically brown rot damage, such as *Coniophora cerabella,* might develop in the softwood sections, whilst a white rot, such as *Polystictus versicolor,* would be more likely to originate in hardwood sections, although the rot may subsequently spread beyond the wood components in which it initially develops. In Britain the treatment of external joinery (millwork) is recommended and is, in fact, a requirement in the case of private houses built under the National House-Builders Registration Council scheme. Similar regulations or recommendations apply in many other countries, as in the Canadian standard for wood windows. Finally, in situations where fungal decay or borer attack is sure to develop, the risk is considered to represent a severe hazard, and this applies to all wood in ground contact.

Alternatively preservative types can be used as a basis for classification, a

Preservative classification

system that is widely used throughout the world. For example, British Standard 1282 includes three types of preservative. Type TO or Tar Oil comprises distillates from coal tar, including creosote. Type WB, or water borne, includes Wolman salts of the fluor-chrome type and also the copper-chrome-arsenic or CCA salts which currently dominate this market. The boron diffusion process for green timber, or Timborising, is also water borne but is usually considered to be a special case, as is the use of aqueous solutions of sodium pentachlorophenate in sapstain control and aqueous emulsions of insecticides in pinhole and Powder Post beetle control. Type OS refers to organic solvent formulations, usually consisting of light petroleum solutions of pentachlorophenol, naphthenates of copper or zinc, chlorinated naphthalenes, organotin compounds and many other less important compounds including contact insecticides. In some areas, particularly Scandinavia, the organic solvent formulations are often decorative, almost intermediate between a preservative and a paint but achieving only a rather doubtful preservative function.

Treatment classification

A third form of classification can be based upon the method of application of the preservative. Full-cell and empty-cell pressure impregnation methods are traditionally used for the application of creosote to wood for service in severe hazard situations but full-cell impregnation is also used for treatment with the well established fixed water borne salt preservatives. Pentachlorophenol in heavy oil has been used as a replacement for creosote in the United states of America but organic solvent solutions are now generally uneconomic unless a high degree of solvent recovery can be achieved. This is, of course, a comment that must be explained in greater detail. The full-cell process, also known as the Bethell process after the name of its inventor, involves placing the wood in a treatment cylinder which is then sealed. An initial vacuum is then produced by pumping, and maintained for from 15 minutes to several hours, depending upon the permeability of the wood and the sizes to be treated. The preservative is then introduced whilst

Plate 3.1

A small impregnation plant installed in a wood yard, showing the baskets on trolleys that are used to load wood into the cylinder. In larger plants the wood is loaded using railway bogies. *(Wykamol Ltd. and Anticimexbolagen)*

Figure 3.6 **Bethell full-cell cycle.**

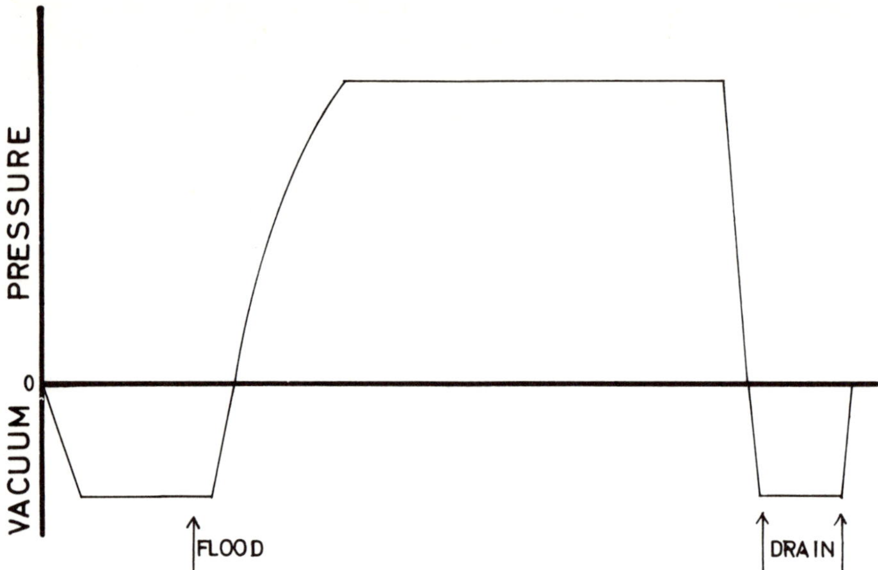

maintaining the vacuum and, when the cylinder is flooded, the vacuum is released and a pressure produced by pumping. This pressure is also maintained for a period depending upon the sizes and properties of the wood to be treated, but is typically for from 1 to 5 hours. The pressure is then released and the preservative removed from the cylinder, followed by a short final vacuum to remove some preservative from the surface of the wood in order to minimise subsequent dripping and bleeding during service.

In empty-cell processes the treatment cycle is modified so that some of the preservative is recovered. In the Rüping process the cycle commences with the application of air pressure to the wood before introducing the preservative. A higher pressure is then applied in order to force the preservative into the wood on top of the trapped air. Finally, the pressure is released, permitting the trapped air to expand and force preservative out of the cell spaces and vessels, a final vacuum being applied in order to

Figure 3.7 **Rüping empty-cell cycle.**

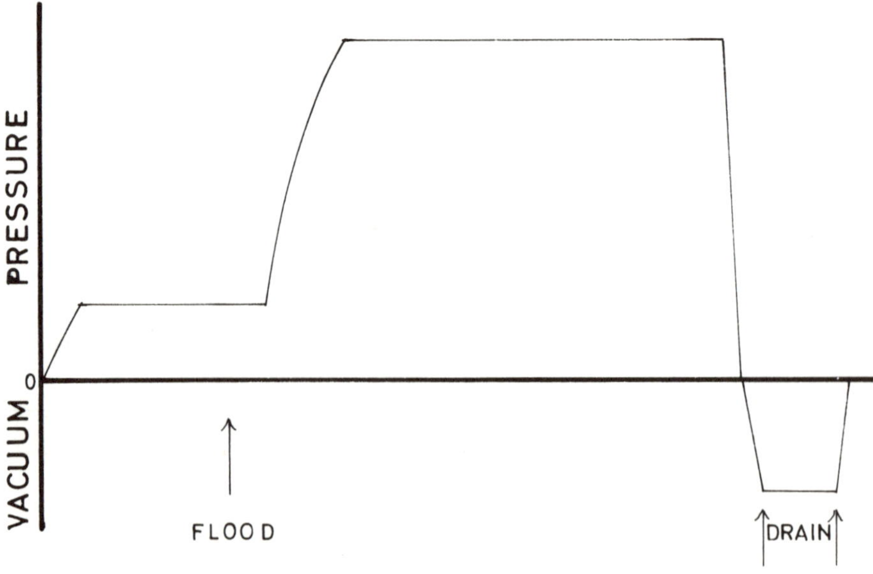

encourage this expansion and thus the recovery of the preservative from within the wood. In this way perhaps sixty per cent of the preservative can be recovered, yet the wood elements are coated with the preservative and thus treated in a thoroughly adequate manner, although with a low toxicity preservative such as creosote the empty-cell process results in a treatment which is not quite so durable as that obtained with the full-cell process. In the Lowry process the initial air pressure stage is omitted and the preservative is introduced into the wood under pressure, compressing the air that is naturally present at atmospheric pressure. Generally a more intense final vacuum is used for a longer period than in the Rüping process in order to assist the recovery of preservative, but recovery seldom exceeds about 50% of the gross absorption.

In the double-vacuum process an initial vacuum is drawn followed by introduction of the preservative. The vacuum is then released and the preservative forced into the wood under atmospheric pressure. Finally, a more intense vacuum is drawn in order to encourage the maximum recovery of preservative. In view of the limited pressure that is applied this process is only considered to be suitable for use with low viscosity organic solvent preservatives applied to reasonably permeable woods which will only be exposed to a moderate hazard. This process is widely used, particularly in Europe, for the treatment of external joinery (millwork) such as window frames for which the use of comparatively expensive organic solvent preservatives is justified; creosote cannot be used because of the dirty finish and water borne preservatives cannot be used because of the distortion that they introduce in the treated components. Unfortunately treatment is comparatively expensive because of the high cost of organic solvents, but an alternative process utilises highly volatile solvents which can be recovered as

Plate 3.2

The impregnation plant shown in the previous plate. This plant can be used for full-cell, empty-cell or double vacuum treatments using organic solvent or water borne preservatives. The preservative is stored in a reservoir under the cylinder, but in large plants there is usually a separate tank. The advantage of this plant is its simple installation. Capacity can be increased by installing a second or third plant fed by the same basket trolley system whereas with larger plants the complete installation must be replaced. *(Wykamol Ltd. and Anticimexbolagen)*

a gas. In the Cellon or Drilon process the preservative is formulated in a liquified butane. Normal treatment cycles can be adopted but the final vacuum is extended for a sufficient time to enable the solvent to be recovered in vapour form. This process suffers from the disadvantage that considerable time and energy is expended in recovering the solvent, and butane is highly inflammable so that the plant must be flushed with nitrogen in order to ensure safety. This serious disadvantage is avoided in the Dow process in which the butane is replaced by methylene chloride.

Figure 3.8 **Double-vacuum cycle.**

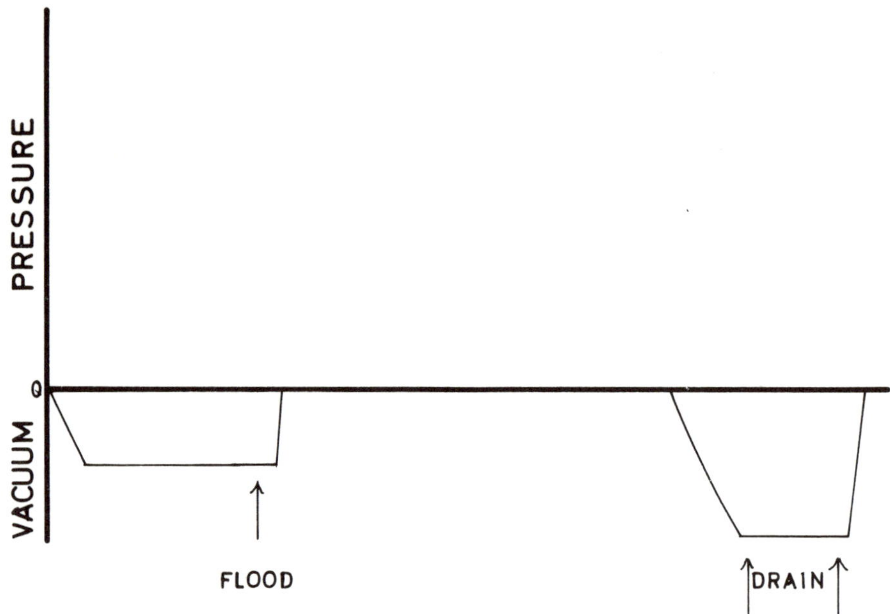

Generally these more sophisticated methods are employed for the treatment of construction timber but, where only a moderate decay hazard is involved, it is perfectly realistic to adopt a superficial treatment method. The use of a paintbrush to apply preservative is generally very inefficient, even when attempts are made to flood the preservative onto the wood. Spray application is far more reliable, particularly if care is taken to ensure that a "flood" application is made, sufficient to flow over the surface but not to excess so that dripping is largely avoided. Deluging is a term that is often applied to treatment by spray tunnel, a system that usually suffers from the disadvantage that it is difficult to achieve adequate loading of preservative on end-grain surfaces. Dipping is perhaps the simplest method for treating wood, consisting of immersion in a tank of preservative for a specified period of time. Prolonged immersion is usually described as steeping. All these non-pressure methods of application are best employed in conjunction with penetrating organic solvent preservatives, the preservative results varying from poor from brushing to very reliable from steeping.

Hot-and-cold treatment

There are several other important methods of treatment. The hot-and-cold open tank method consists in its simplest form of a drum or tank filled with preservative over a fire. The wood to be treated, such as fence posts, is placed in the tank which is then heated, expanding the air in the wood. The fire is then permitted to burn out and the progressive cooling causes the air in the wood to contract, drawing in preservative. In more sophisticated plants the wood is immersed first in a tank of hot preservative, followed by transfer to a tank of cold preservative, but the principle remains the same. In the boron

Plate 3.3

A deluging plant, or spray tunnel, used to treat joinery (millwork). Similar equipment can be used for anti-stain treatment. *(Wykamol Ltd.)*

Plate 3.4

A simple dip tank fitted with a draining board. The tank is assembled from bolted sections so that the length can be adjusted and it is also easily transported so that wood can be treated in this way on building sites after all cutting and woodworking is complete. *(Wykamol Ltd.)*

Diffusion treatment

diffusion treatment freshly converted green wood, with a moisture content in excess of 50%, is dipped in a hot concentrated solution of a borate. The wood is then close stacked and wrapped with sheeting in order to prevent evaporation of the moisture. These conditions are maintained for several months, depending upon the cross-section of the wood, so that the borate is able to diffuse through the moisture throughout the wood. This system is sometimes used with fluoride preservatives but it will be appreciated that, in both cases, diffusion is only achieved because of the water solubility of the preservative system and this necessarily results in a treatment that is also susceptible to removal by leaching in its ultimate use situation. These methods are therefore unsuitable for use in ground contact or other conditions where there is a danger of persistent leaching and where reliably fixed preservatives, such as the copper-chrome-arsenic (CCA) salts, must be applied by vacuum-pressure impregnation.

Retention and penetration

The effectiveness of a toxic preservative treatment depends upon the amount of preservative that is deposited within the wood; this is termed the net retention and must obviously be sufficient to prevent the attack of organisms that are likely to be encountered in the particular use situation. In addition, effectiveness depends upon the depth of penetration. In theory it may appear that a superficial coating of preservative is adequate but in practice significant penetration is essential. Some fungi and borers are able to tolerate limited deposits of preservative which they detoxify, and it is essential that the preservative retention and penetration is adequate in order to prevent damage to the treated zone and thus exposure of the unprotected wood beneath. In addition, wood that is exposed to the weather will be subject to severe movement stress through alternations of dry and wet conditions, frequently resulting in the development of severe splits, particularly in woods with large or moderate movements, which may penetrate through the treated zone, trapping water from rainfall and encouraging decay to become established within the unprotected wood beneath the treated zone. Clearly adequate penetration is essential, and deepest penetration is required when wood is exposed to the weather. In the case of Baltic redwood, *Pinus sylvestris,* telegraph and power transmission poles the heartwood is relatively durable and is surrounded by sapwood which is sufficiently permeable to permit complete treatment with creosote or a water borne salt preservative applied by vacuum-pressure impregnation methods. In Douglas fir, *Pseudotsuga menziesii (taxifolia, douglasii),* the heartwood is similarly durable but the sapwood is resistant to impregnation. However, the sapwood is most permeable in the longitudinal direction, whilst the permeability in the tangential direction is far higher than in the radial direction. It is therefore normal practice to incise the poles prior to treatment by using knives orientated in the longitudinal direction, to open up particularly the tangential routes. In spruce, *Picea* spp., neither the sapwood nor the heartwood are naturally durable and neither are readily penetrated by preservatives. For these reasons spruce is generally unsuitable for transmission poles, although it is sometimes used after incising and with a long impregnation period to encourage deep end-grain penetration of the heartwood.

Incising

Unfortunately spruce grows rather more rapidly than most species of pine in the temperate and coniferous regions, and considerable efforts are being made to improve its utilisation by developing preservative treatments capable of deeper penetration. At the present time only borate or fluoride diffusion treatments will achieve complete impregnation but a system for incising sawnwood has been developed by the Princes Risborough laboratory in England which enables the preservative to penetrate to the depth of the incising pattern, usually about $\frac{1}{4}$ in. (6 mm.). Whilst this is far from perfect, it is at least far more efficient than any alternative method. Knives are not the only means for cutting wood, and high-energy liquid jets can be used to make a similar incising pattern, the depth of cut depending upon the pressure used. A preservative formulation can be used as the cutting fluid, enabling the process to achieve both incising and treatment in a single operation. The energy costs for operating the system are rather high if it is required to achieve the normal absorption of preservative within the wood but it seems possible that the method may be developed in future around concentrated preservatives which can be applied at comparatively low absorptions, spreading within the wood to give the required distribution and retention of toxicants.

High-energy jets

Sap displacement method

One further preservative application system remains to be described; the sap displacement method. In the Boucherie method a cap is attached to the butt of a log immediately after felling and preservative is introduced at a low pressure, normally from a header tank, so that the sap is displaced and replaced with the preservative solution. In its original form copper sulphate solution was used as the preservative but it is largely unfixed and the treated wood is generally considered to be unsuitable for use in leaching conditions, including ground contact. Fixed multi-salt preservatives have been used more recently, although some types tend to fix too rapidly so that they are filtered out as the preservative passes up the log, the butt end being far more heavily treated than the top or crown end. Thus the copper-chrome-boron and the fluor-chrome-arsenic-phenol treatments are far more suitable for application by this method than the copper-chrome-arsenic preservatives that are so widely applied by vacuum-pressure impregnation methods. In some adaptations of the method the butt end of the log is immersed in the treatment fluid and a vacuum cap attached to the top end. Although Boucherie first patented this process in 1834 it still attracts considerable interest, particularly in under-developed countries where it is perhaps the ideal process for the preservation of logs for use as poles and piles. In fact, the process is particularly suitable for wood that will be used in the ground as only the sapwood can be treated by this method. Modern developments have concentrated on producing improved cap systems and preservative formulations that will distribute evenly yet ultimately fix within the wood and become resistant to leaching.

Factors influencing penetration

The penetration of preservatives applied by superficial methods, such as brushing, spraying and immersion, will depend largely upon the length of time that the wood is in contact with the preservative coupled with the rate of penetration. Non-polar organic solvent formulations achieve far greater penetration than polar water borne formulations but in practice the penetration from brushing or spraying will depend upon the loading of preservative that can be achieved on the surface of the wood, and this will depend in turn upon the viscocity and the volatility of the formulation. In the case of immersion the penetration will depend principally upon the time of immersion, although viscosity and surface tension will be very important; high viscosity creosote will penetrate to only a limited extent compared with low viscosity organic solvent formulations. Viscosity is equally important in the case of vacuum-pressure impregnation; indeed creosote is frequently heated to lower its viscosity and improve its penetrating power. In other respects the penetration will be dependent upon the permeability of the wood, coupled with the intensity of the vacuum and pressure stages of the treatment cycle.

In summary, pressure impregnation with creosote is normally used for the treatment of heavy structural components in ground contact or marine conditions, such as posts, transmission poles, piles and railway sleepers (ties). Pressure impregnation with fixed water borne salts is used for similar purposes, except that the creosote achieves greater dimensional stability and thus resists the serious splitting that may result from alternations of wet and dry conditions and which may penetrate through the treated zone. Pressure impregnation with fixed water borne salts is also frequently used for the treatment of sawn carcassing (structural framing) in buildings, although organic solvent preservatives applied by double-vacuum or low pressure impregnation methods, or by immersion, are being used to an increasing extent and are used almost exclusively for joinery (millwork) where water borne preservatives are unacceptable because of the distortion that they may produce. The sap displacement methods are of greatest importance in relatively undeveloped areas where their unsophisticated equipment can be used to produce reliably treated logs for use as poles and piles. In theory diffusion treatment has considerable advantages over all other methods as a means for treating general building wood, as it achieves penetration throughout the cross-section so that the wood can be resawn or worked without exposing untreated wood. In fact the present treatments applied by diffusion suffer from the distinct disadvantage that they are susceptible to loss by leaching and this severely limits their usefulness.

It is unrealistic to attempt a description of the several thousand wood preservative formulations in use in various parts of the world; the reader is referred to the companion volume "Wood Preservation" for this information. In principle wood preservation enables the user to select a wood that is economic, readily available and particularly suitable for the desired use; preservation will then provide the durability that is necessary. Whilst there is always a danger that all but the most durable woods will suffer from fungal decay or borer attack, the likelihood of damage clearly depends upon the use situation, and preservation cannot be universally justified. Obviously there are situations, such as ground contact use, where preservation is essential in virtually all parts of the world. There are other situations where local pests, such as dry wood termites or House Longhorn beetle, make an appropriate preservation treatment equally essential and perhaps necessary to meet local building regulations. In other situations, such as the general treatment of structural wood and external joinery (millwork) in temperate climates, preservation is desirable but not essential; careful maintenance will be sufficient to avoid damage and the treatment is only an insurance against accident or neglect. Finally, there are those situations where the risk of damage is slight and the wood relatively durable, such as where tropical hardwoods, virtually entirely heartwood, are used in the manufacture of furniture.

Remedial treatment

This brief description of wood preservation would be incomplete without a description of remedial treatment, or the application of preservatives in order to eradicate established borer attack or fungal decay. Remedial treatment commences when the damage is observed or suspected, and the wisest course is then to instruct a specialist company to make an inspection. A considerable amount of knowledge and experience of both structures and the wood destroying organisms, insects and fungi, is essential if the inspection is to be competent. For example, both termites and House Longhorn beetles can cause very considerable damage before there is any external evidence of their activity, and the dry rot fungus can spread though plaster, brickwork, masonry and concrete in its search for nourishment and more distant pieces of wood within the structure. The inspection must detect the extent of the damage or, alternatively, suggest the areas which should be opened up to permit a more detailed inspection. The first task of the treatment operatives is to expose the full extent of the damage in order to decide whether the affected components should be replaced with adequately preserved wood or whether treatment to eradicate the fungal infection or insect attack will be sufficient. This "opening up" process is perhaps the most important aspect of conscientious remedial treatment as it largely ensures that concealed damage is detected. Preservative treatment then follows, although fungal decay can generally be attributed to a fault in the structure, perhaps accident, negligence or a design defect, permitting wood to become wet, and this fault must be corrected as part of the treatment. It is not sufficient to treat only the wood that is visibly affected; adjacent wood may already be infected by the fungus concerned and, in the case of an insect infestation, an attack in one wood component will clearly indicate the danger faced by all others in the structure.

Remedial treatment must be both eradicant and preservative, so that the preservative formulations are often pretreatment formulations with additional eradicant components. Spray treatment is normally used in conjunction with organic solvent formulations which are both more penetrating than water borne types and also free from the staining that often occurs when, for example, a roof treatment accidentally soaks into a plaster ceiling.

Plate 3.5

Remedial treatment. Spraying woodwork with an organic solvent formulation, to eradicate a beetle infestation and to preserve the wood against future attacks. *(Richardson & Starling Ltd.)*

Occasionally holes are drilled into beams and other large wood components to facilitate treatment by pressure injection; large beams in ancient buildings are often in contact with damp masonry either end and therefore suffer from internal decay through the absorption of water by the permeable end-grain. Brickwork and masonry infected by the dry rot fungus are also drilled to permit proper sterilization, usually with an aqueous formulation as the presence of fungus in the wall indicates dampness which might obstruct the spread of an organic solvent preservative. Whilst the stripping of all damaged wood, coupled with spray application of preservatives, is the most reliable method for remedial treatment, there are less sophisticated methods used in many parts of the world. In Britain the use of contact insecticide smokes has been recommended as a method for eradicating insect borers such as the Death Watch beetle. In fact the smokes do not penetrate into the wood and leave a deposit, largely on upper horizontal surfaces, which may kill emerging adult beetles and thus prevent subsequent egg laying. Unfortunately the contact insecticides that are used, such as Lindane, Dieldrin and DDT, have little persistence when finely dispersed on the surface of wood in this way and treatment must be repeated at intervals for a period of perhaps eight to twelve years, to permit all larvae within the wood to pupate and emerge as adult insects which will be killed by the insecticidal deposit. Even prolonged treatment can fail to control Death Watch beetle in the interior of large beams; this insect is always associated with fungal decay and beams can be hollowed out, as previously described, if their ends are built into damp masonry. Toxic gases, particularly methyl bromide, are also used to eradicate insect infestations, but they have distinct disadvantages. The structure must be enclosed within an impervious sheet during treatment and must subsequently be well ventilated to thoroughly remove the toxic gas. In addition, the gas will only kill insects within the wood and the treatment has no preservative action to prevent subsequent reinfestation.

3.3 Water repellents

Changes in the moisture content of wood can introduce two different defects. Changes in moisture content up to fibre saturation point invariably involve movement, shrinkage with drying and swelling with wetting. Although it is normal to dry wood to a moisture content equivalent to the average atmospheric relative humidity anticipated in use, it is common to encounter movement problems. Faults such as gaps appearing between floor blocks or boards are due to the wood drying after installation, either through inadequate kilning or perhaps rewetting between kilning and installation. A door or drawer jammed in humid weather may be exceedingly slack under drier conditions. Frames which introduce an end-grain surface in contact with side-grain will inevitably result in cracking of any surface coating system. In other situations the cross-sectional movement may become apparent as warping through twisted grain effects. The obvious solution to all these problems is to use only wood with low movement but this is not always realistic. The alternative is to impregnate the wood with chemicals which induce stabilisation. Unfortunately these processes, which are described in more detail in the book "Wood Preservation", are also frequently unrealistic because of the difficulty of achieving impregnation, a problem that has already been discussed in connection with normal wood preservation.

Surface coatings

The only obvious solution is to enclose the wood within a protective film in order to stabilise this moisture content. Paint and varnish coatings will act in this way, provided they completely cover the wood and provided they are not damaged in any way. Unfortunately, whilst these coatings give good protection against rainfall, they are unable to prevent moisture content changes resulting from slow seasonal changes in atmospheric relative humidity. As a result the painted wood will shrink or swell with changes in relative humidity, causing the surface coating to fracture wherever a joint involves stable side-grain in contact with unstable end-grain. Rain is absorbed by capillarity into the crack, yet the remaining paint coating restricts evaporation so that the moisture content steadily increases until fungal decay is sure to occur if the wood is non-durable. It is frequently suggested that preservation provides a simple solution to this problem, but this ignores the fact that the water also damages the paint coating. It is explained in Appendix 2 that wood is a hygroscopic material, covered with hydroxyl groups which have a strong affinity with water, so that penetrating water will tend to coat the wood elements, displacing paint and varnish coatings. This failure is known as preferential wetting and is responsible for blistering and peeling in paintwork, and the loss of transparency in varnishes.

Water repellents

The best solution to both the decay and preferential wetting problem appears to be the treatment of external joinery (millwork) and cladding before painting with a formulation that is both a preservative and a water repellent. In fact, the term water repellent refers specifically to treatments which coat the pores of a structural material, reversing the angle of contact so that capillary absorption of water is destroyed and the surface becomes "water repellent". Whilst the prevention of liquid water absorption is clearly an advantage, and whilst a suitable treatment of this type may also serve to prevent preferential wetting, a reduction in permeability is also desirable in order to oppose moisture content changes through fluctuations in atmospheric relative humidity. Resins are therefore frequently included in formulations of this type in order to improve the resistance to water vapour transfer. Whilst many water repellent preservatives are designed specifically for use as pretreatments prior to painting or varnishing, perhaps in place of conventional priming treatments, other water repellent preservatives are designed as complete maintenance treatments, frequently serving a decorative as well as protective function; these decorative preservatives are particularly popular in Scandinavian countries. These two types of water repellent preservatives are not necessarily similar; the first type must be compatible with subsequent paint or varnish coatings, whilst the second type must clearly have good resistance to weathering. A further type of water repellent preservative is also available, consisting of a normal water borne multi-salt preservative with the addition of emulsified water repellent components. The purpose of this formulation is to reduce moisture content fluctuations in wood exposed to the weather, in order to reduce the stresses that cause serious splitting damage; these formulations are an attempt to make water borne preservatives as reliable as creosote in some severe hazard situations.

3.4 Fire retardants

Fire resistance

Fire will only occur when combustible material is subjected to sufficient heat in the presence of oxygen. In the absence of any one of these three components ignition cannot occur. Within a building it is necessary to have fire resistance; the ability of the building components to withstand fire

Spread of flame

penetration so that an accidental fire remains isolated for a sufficient period to permit the occupants to escape and to give reasonable time for the fire service to arrive and prevent further damage. Fire resistance is most important where a building is divided into relatively small compartments, but in large open spaces, such as corridors, stairwells, and roof spaces, it is more important to prevent the rapid spread of flames across surfaces. Fire resistance is best achieved in a building by using an adequate thickness of wood in construction and avoiding any minor imperfection which will permit the fire to penetrate through a partition or fire barrier. Clearly doors should be tightly fitting, although one solution to this problem is to insert a special strip in the edges of the doors or the door frames. This strip is composed of an intumescent material which swells when subjected to fire and thus seals the gaps. The fire resistance of a structure is also improved if the wood is not combustible, and this applies equally to preventing the spread of flame in large spaces in buildings.

Fire retardants

There are two principal systems which are used to increase the fire resistance of wood structures and to reduce the surface spread of flame. The simplest system is perhaps the application of a special intumescent paint coating. When exposed to fire the coating foams, introducing a thermal insulating barrier which prevents rapid increase in the temperature of the wood beneath, and which contains fire retarding gases which further restrict the possibility of wood ignition. Intumescent coatings are only realistic where they can be applied as part of a decorative system. The alternative is the impregnation of the wood with fire retardant salts. These prevent normal ignition, although some surface charring may occur to a very shallow depth. These impregnation treatments are particularly suitable for structural sawn wood. Most proprietary products are hygroscopic and susceptible to loss by leaching, so that they cannot be used, for example, for the treatment of external cladding; if they are unprotected they will be lost by leaching, and surface coating systems are inefficient as they fail by preferential wetting due to the hygroscopic treatment absorbing water.

This is necessarily a very brief description of the use of fire retardants. More detailed information is included in the companion volume "Wood Preservation".

4. Wood Utilisation

4.1 Basic principles

Wood in tension

Wood can only be efficiently utilised if there is an intelligent understanding of its properties. Wood has very high strength in compression along the grain so that a wood post is a very efficient form of support. Wood is also light and strong as a beam but it must be relatively free from defects, particularly the presence of knots on the lower edge which is in tension; the stress grading system of sorting wood is designed to ensure that individual pieces are capable of meeting the design requirements in a structure. Wood is very weak in tension across the grain, yet very strong in tension along the grain. Unfortunately this means that considerable difficulties are encountered in obtaining fixings to a piece of wood under tension. If a hole is drilled through the wood the first effect is to reduce the cross-section area and therefore the load-carrying capacity. The second effect in tension is that the loading falls on the side of the hole closest to the end of the piece, so that the piece of wood supporting the load is only connected to the remainder of the pieces by cross-grain cohesion. In fact, the sheer strength of wood is very poor in the cross-grain direction and there is a danger that the loaded plug of wood will pull out from the end of the piece. This danger is considerably reduced if the load can be spread across the entire width and, if possible, the depth of the piece by clamping to the supporting member, but there is still a danger that fluctuating moisture content will cause shrinkage and swelling which will eventually result in a loosening of the joint. This is generally overcome by bolting the members in tension together with a connecter between them, consisting of a plate with a large number of spikes which penetrate the wood on either side of the joint. For lighter loads gangnails or toothed plate connectors can be used, consisting of plates with a large number of spikes which can be hammered over the joint between two members in, for example, a light roof truss.

Figure 4.1 Bulldog connector.

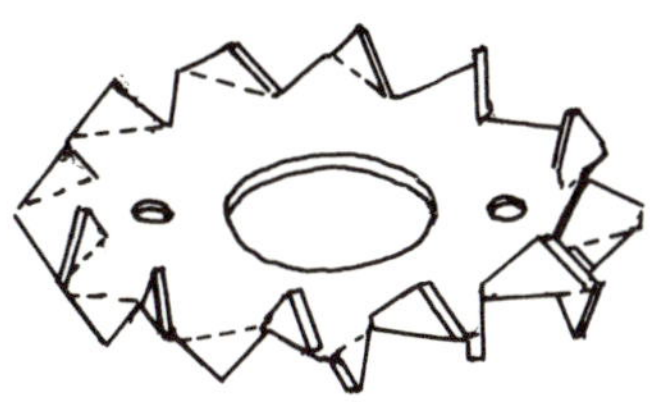

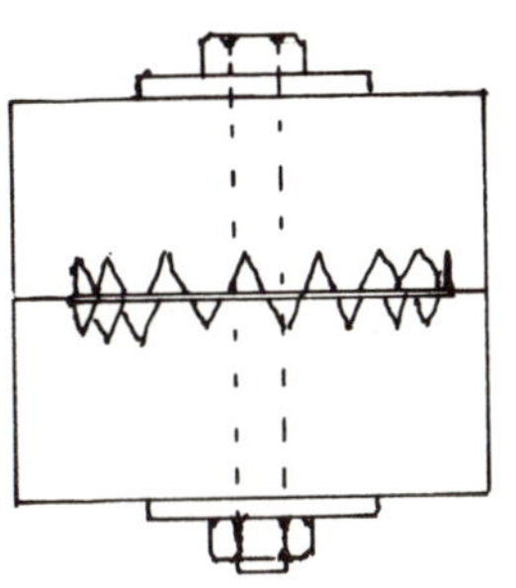

Wood in compression

It will be appreciated that, whilst it is necessary to spread a load evenly when a piece of wood is in tension, the situation is somewhat simpler in compression. A beam must simply rest squarely on a post and it is only necessary to provide a low strength fixing to ensure that the post and beam remain in their proper relationship. One of the traditional methods for achieving this jointing is the mortice and tenon where one piece is cut to provide a tongue, the tenon, and the other is cut to provide a slot, the mortice. In post and beam construction the tenon is cut in the post with the mortice in the beam so that the maximum cross-section of beam rests upon the maximum cross-section of post; a serious criticism of mortice and tenon joints is the loss of cross-section that occurs and thus the loss of strength. As far as the post is concerned the loss of bearing area from a thin tenon is relatively unimportant but for the beam the loss of cross-section area by the mortice is very important, and this method of construction is only realistic because a beam is most lightly stressed at its ends where the mortice is cut and most heavily stressed at the mid-point which must therefore be free from defects. Unfortunately some form of bracing must be provided to ensure that the post remains upright and the beam horizontal. Traditionally a diagonal brace is provided with a diagonal mortice and tenon joint at either end but this has the disadvantage that, as the beam becomes loaded, the brace will tend to bend the post at precisely the point where it is already weakened by the presence of the mortice in which the tenon of the brace is inserted. Simplified systems, such as sheets of plywood glued and nailed to the sides of the post and beam, provide a bracing which is efficient but which does not reduce the strength of the principal structural members.

Panelling

These are the engineering principles involved in the structural use of wood but there are non-structural aspects which must be recognised. Perhaps the most important of these is the high movement in the cross-grain direction with changes in moisture content. It will be appreciated that, in the case of structural members such as posts, beams and braces, the stable longitudinal dimension is of most importance but the cross-grain dimension becomes of importance whenever wood is used for panelling. If boarding is to be employed for flooring or any other panelling application it must be selected for low or medium movement, and woods with large movement must be considered to be entirely unsuitable. Alternatively, plywood and perhaps particle and fibre boards offer the most realistic alternative.

Strength, texture and durability

A softwood generally has very high compressive strength along the grain and bending strength across the grain, so that it is particularly suitable for general structural uses as posts, beams and similarly loaded components. However, it possesses very low compressive and tensile strength across the grain and, where this is an important factor, some of the hardwoods, particularly those with interlocking grain, are undoubtedly more suitable. Obviously a considerable amount of selection is necessary to ensure that the most suitable wood is used for a particular purpose, so that a coarse-textured wood might be perfectly suitable in the structural engineering sense yet entirely unsuitable for the production of joinery (millwork) or furniture where a fine finish is required. In addition, naturally durable or adequately preserved wood must always be used where there is a danger of fungal decay or insect attack, such as whenever wood is used in ground contact or in areas where there is a termite or House Longhorn beetle problem.

Adhesives and finishes

It is useless selecting the correct wood and then utilising it in the wrong way. There are, for example, a variety of adhesives that are available but some have inadequate water resistance whilst others are too rigid and are unable to accept the cross-grain movement that occurs with changes in relative humidity in the surrounding atmosphere. Indeed, a separate book could be written about adhesives but the best procedure in each part of the world is to accept the recommendations of the local standards organization. A further book could be written about paint and varnish finishes for wood, and again there are particular types of finish that are more suitable, or perhaps just more popular, in certain parts of the world. In general a purchaser receives the goods for which he pays; if the cost is skimped this usually means inferior materials or workmanship and a consequent loss in reliability and durability. It is not sufficient, however, to select expensive, recommended or specified adhesive or finish systems as their performance will also depend upon their compatibility with the wood and this varies very widely with species. The most important factors are the permeability of the wood, its stability and the presence of interfering extractives.

4.2 Building structures

Primitive buildings

Wood is used in buildings as the structural support or as the weather-proof cladding. In the simplest structures only post and beam components are involved and these are perhaps best illustrated by the buildings that are still observed in undeveloped tropical countries. Generally these buildings consist of a series of posts to which simple beams are tied to provide a sloping roof. There is no attempt at triangulation to ensure rigidity which is instead obtained by deeply embedding the posts in the ground. Generally the beams are fixed to the posts by cords, perhaps natural vines, and when a ridge is introduced it is generally supported on longer posts, so that a ridged building consists essentially of two lean-to buildings constructed together. Sometimes suspended floors are used, again normally fixed to the poles by cords. Small diameter poles are perhaps bound together to form the floor, and the roof is covered with large leaves or thatch to give protection from rainfall. The wall panels are then completed with poles or mats. In other cases the posts are replaced with solid walls constructed from stone or clay.

The next development in building probably resulted from the introduction of squaring. Initially logs were simply trimmed to a square shape but more sophisticated methods of conversion were progressively introduced, such as splitting suitable woods such as chestnut and ultimately sawing so that boards became possible. In many areas a frame construction was adopted for wooden buildings so that each wall had a lower horizontal plate or ground-sill supporting the principal posts. The posts in turn supported the somer or wall plate at their top and bressumers at intermediate levels. These components completed the basic wall frame, and beams were fitted between opposite posts to support the floor joists and, in turn, the floorboards. Opposite posts at wall plate level were joined by tie-beams and the entire system suitably braced.

This form of frame building normally supported a pitched roof. The main load bearing elements were the trussed or principal rafters running between opposite posts and incorporating the tie-beams. In the simplest form of rafter a king post from the centre of the tie-beam was provided to support the ridge piece, and the principal rafters then ran from the ridge piece to the wall plate. In a normal domestic roof a purlin would be positioned mid-way down each rafter in order to space the rafters uniformly and to resist any deflection due to the tile loading and any additional snow or wind loading. As a load carrying member the purlin obviously required support and this was usually provided by braces running to the base of the king post together with a horizontal collar between opposite purlins.

The king post system is comparatively inefficient because of the loading that it may place on the centre of the tie-beam. An alternative is to provide the purlins with a collar, sometimes known as a wind-beam, and then support each end of the collar, and thus the purlins, on queen posts running down to the tie-beam. Even in this form of construction a small king post could be provided from the centre of the collar to support the ridge, although braces from the centre of the collar to the adjacent rafters might be more normal. Intermediate rafters would be supported on the ridge piece, the purlin and the wall plate, and in turn the rafters would be covered by horizontal laths to support the tiles, slates or thatch.

The wall panels would be completed with vertical studs, puncheons or quarters and perhaps horizontal cross-quarters. The spaces between these components might be filled with brick or "wattle and daub", a crude plaster

Figure 4.2 **Traditional English wood-frame construction.**

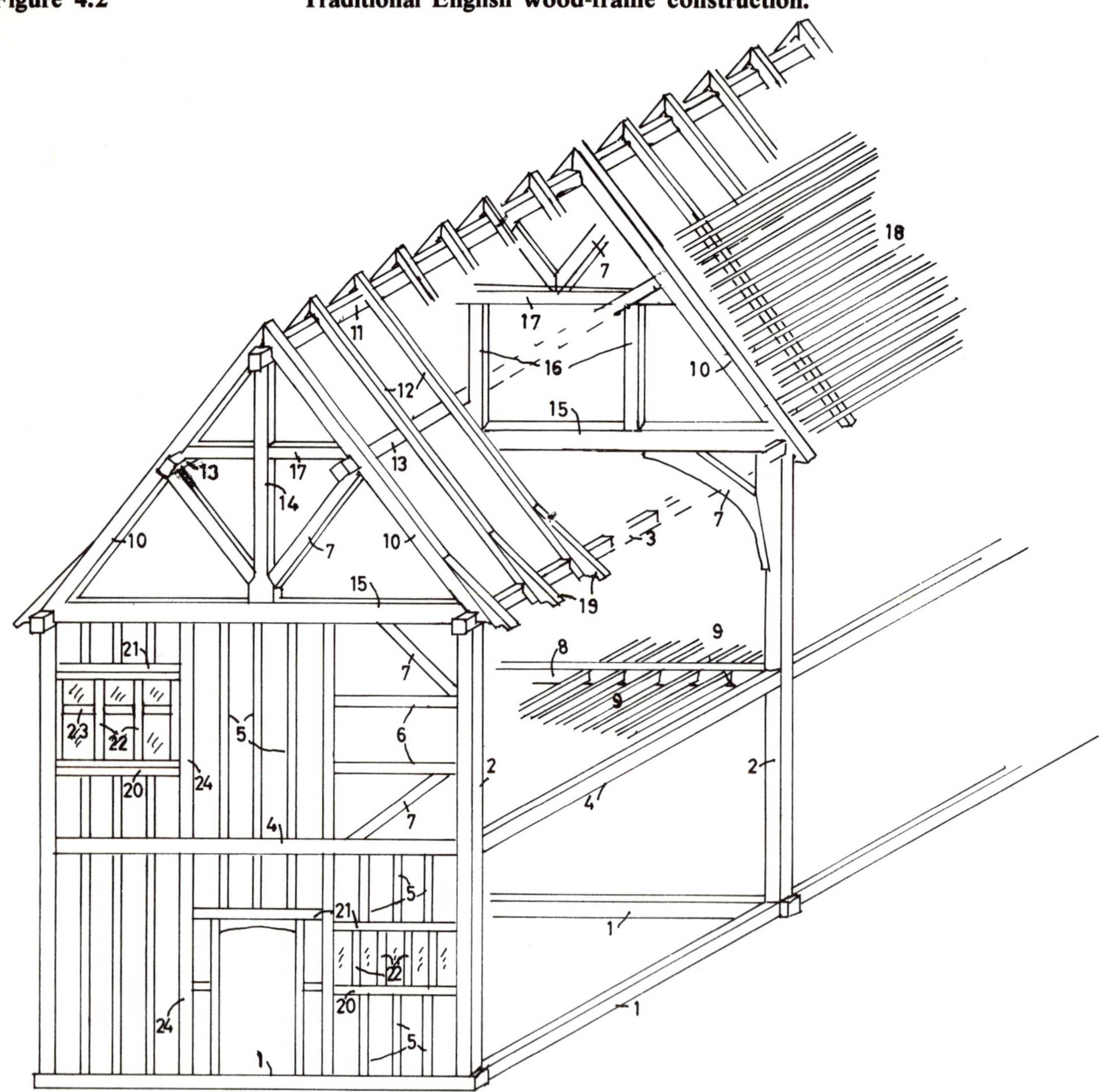

1 = ground sill, 2 = principal posts, 3 = somer, wallplate, 4 = bressumer, 5 = studs, puncheons or quarters, 6 = cross-quarters, 7 = braces, 8 =beams, 9 = joists, 10 = principal rafters, 11 = ridge piece, 12 = common rafters, 13 = purlins, 14 = king post, 15 = tie beam, 16 = queen posts, 17 = collars, wind-beams, 18 = laths, 19 = firrings, 20 = sill, 21 = lintels, 22 = muntins, 23 = transoms, 24 = posts.

Plate 4.1 **Traditional English wood-frame construction.**

mix reinforced with ossier and perhaps straw and horse hair. Alternatively in areas where pit sawing was used the outer surface might be clad with horizontal or vertical boarding and the inner surfaces lined with wainscot panelling. Boards were first prepared by cleaving rather than sawing and the use of boards was traditionally confined to temperate areas producing particularly oak which would be cleft in this way. In other areas the need for a weatherproof wall was solved by using horizontal log construction, as in Scandinavia. Vertical posts were not used and instead the building was assembled using a corner jointing technique as shown in Figure 4.3. The upper log had a concave section to form a snug fit over the lower log, and a slot in each log relieved the movement stresses, reducing the development of splits over the log surface. Often the logs were trimmed to a neater oval section and the roof was completed with thin poles covered with birch bark and finally a turf covering. The Scandinavians did, however, use post construction later in both barns and their beautiful stave churches.

In principle the use of wood has varied very little from the traditional frame plan. The wall framings remain basically the same but the size of the components has been reduced to the minimum that is required. This has been largely possible through the introduction of improved fixings; one reason for

Figure 4.3 **Traditional Scandinavian log construction.**

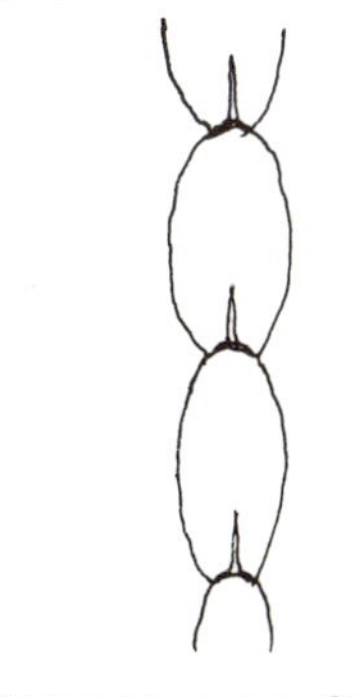

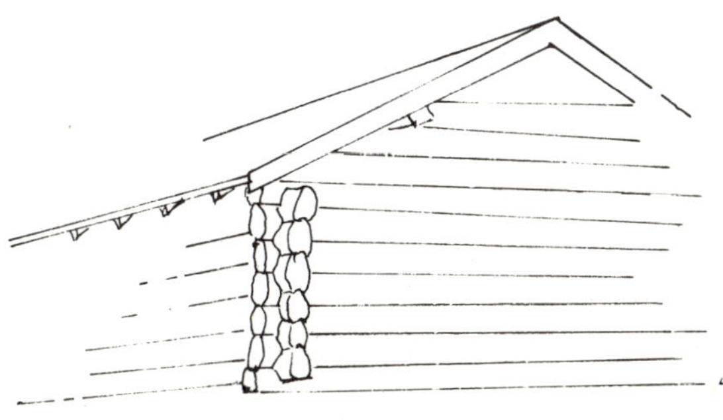

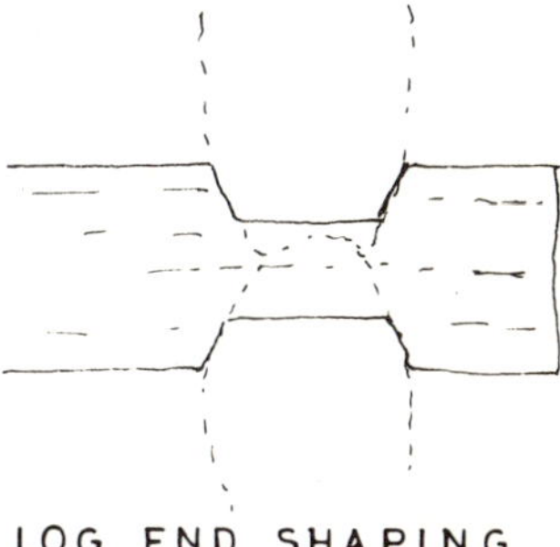

Plate 4.2 Traditional Scandinavian log construction.

Plate 4.3 Models of a modern wood-frame house. *(Council of Forest Industries of British Columbia)*

the use of large dimensioned wood in the past was the loss of strength that occurs wherever the components are jointed using the traditional mortice and tenon techniques. In recent years perhaps the greatest developments have been in roof trusses where designs now eliminate the point loading on beams previously associated with the use of king- and queen-posts. It is impossible to describe all the minor variations in construction that are now adopted; the purpose of this book is to explain the basic principles and to give examples, and it does not attempt to be a comprehensive reference work on construction. However, a number of basic precautions must be mentioned.

Figure 4.4 Trusses.

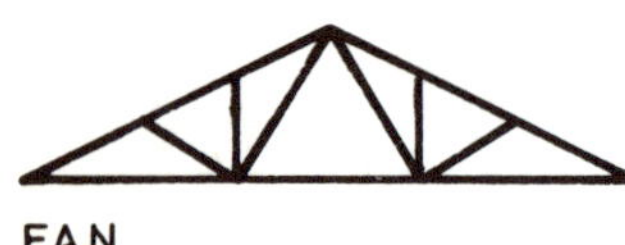

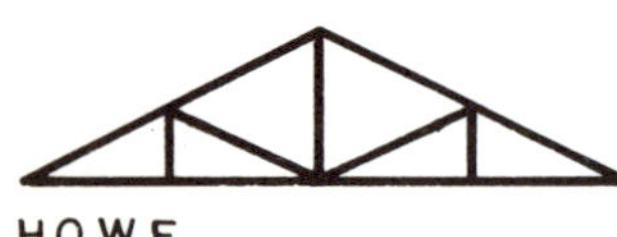

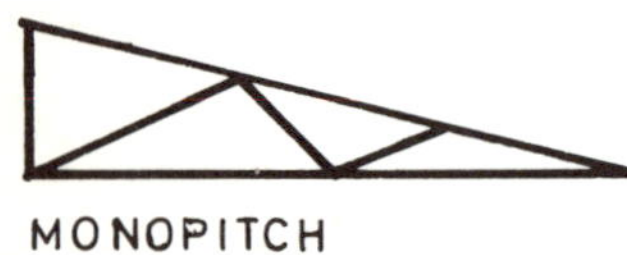

Generally wood in buildings must be completely protected from dampness. The wood must be isolated from contact with the soil by the provision of an impervious damp-proof course. The roof must be designed to either throw the water well clear of the walls or alternatively to collect the water and to channel it into a proper rain water disposal system. Where wood is used as a cladding it must be naturally durable or adequately preserved, although sometimes non-durable wood is used as a cladding, relying for its protection upon a paint system in the same way as external joinery (millwork) such as window and door frames. One particular problem with this form of cladding is the moisture gradient that develops in winter weather between the warm interior of the building and the cold exterior. The interior air has a high moisture holding capacity so that, although its actual moisture content may be high, its relative humidity is low. As this air diffuses towards the external surface it is cooled, reducing its moisture holding capacity so that its relative humidity rises until eventually condensation occurs. If this condensation occurs within non-durable wood cladding it may cause severe decay and, if the cladding is painted, the accumulation of water just below the external surface will cause the paint to lose adhesion and fail by blistering and peeling. The danger of this condensation failure is generally reduced by incorporating a moisture barrier membrane, such as a bitumenised paper, within the walls.

Joinery (millwork)

Joinery (millwork) is particularly prone to failure. The main problem arises at the corner joints where unstable end-grain butts against stable side-grain. Whilst a paint finish on frames will prevent the absorption of incident rain-water, it is unable to completely resist the seasonal changes in relative humidity which will cause the components of the frame to expand and contract in cross-section but not in their longitudinal dimension. As a result of this differential movement the paint at a corner joint splits and subsequent rainfall is then absorbed by capillarity. Unfortunately the remaining paint effectively prevents drying by evaporation, and moisture progressively accumulates until there is a severe danger of decay arising if the wood is non-durable or inadequately preserved. In fact preservation can achieve only half the protection that is necessary; it will prevent decay but it will have no influence over the failure of the paint film through preferential wetting, or the spreading of water over the surface of the wood causing loss of paint adhesion. It is sometimes suggested that the use of a water repellent preservative will prevent this defect but it is probably more truthful to state that it reduces the danger and it is really necessary to consider a fundamental change in construction in order to avoid this faulty joint.

One solution to this problem is to construct the frames from a naturally durable wood with a low movement, such as Douglas fir (Colombian pine) heartwood. Whilst this is technically realistic it is unfortunately economically unsound in view of the enormous amount of joinery used throughout the world and the limited availability of suitable woods. The alternative is to redesign the joints. A significant improvement can be obtained by assembling the corners using a multi-comb joint rather than a simple tenon and mortice as there is then some restraint exerted by one component upon the other. One further and apparently logical improvement is to make the surface joint a mitre so that each component is equally affected. In fact such a joint is not as logical as it might appear as shrinkage in the two components results in the joint pulling open at its inner end. Probably the most successful solution to the problem is a finger joint on a mitre so that the end-grain of each piece reaches the surface as a feather edge which is readily restrained by its attachment to the longitudinal dimension of the adjacent piece of wood. It is probable that the universal introduction of mitre finger joints for joinery assembly would result in the avoidance of this major source of trouble in window and door frames.

Figure 4.5 Fingered corner joint for window frames.

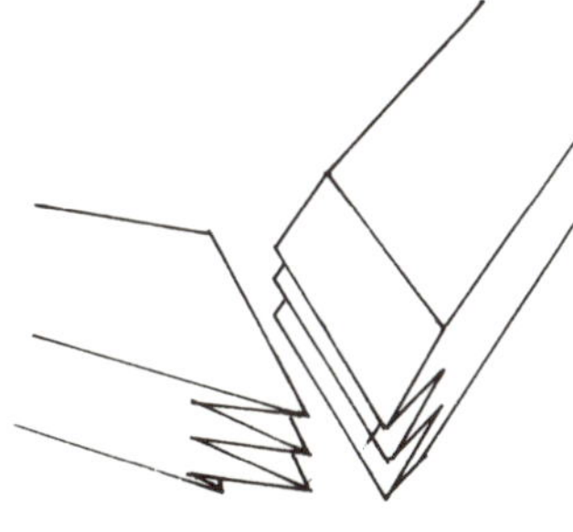

Mention may be made of several recent developments in the use of wood in buildings. In North America preservative treated Douglas fir plywood is now being used successfully in the construction of house foundations, extending building operations to the winter when normal concrete foundations cannot be laid. Throughout the world there is a tendency for factory-finished components to be used, particularly plywood panels finished with a durable coating. Unfortunately it is difficult to ensure that these finishes remain undamaged during installation and, in the case of panels, the system is only

Plate 4.4 North American style frame construction in England. Western red cedar cladding (siding) is painted white. The roof is clad with western red cedar shakes. *(Council of Forest Industries of British Columbia)*

Plate 4.5 Western red cedar cladding on a house near Edmonton, Canada. *(Council of Forest Industries of British Columbia)*

Plate 4.6 Modern wood construction in England. The lightly framed plywood sections used for walls and partitions are the only structural components in these four-floor maisonettes. *(Timber Research and Development Association)*

Plate 4.7 Preservative treated framing and Douglas fir plywood being used in Canada for the construction of house foundations, a system that is now widely employed in North America. *(Council for Forest Industries of British Columbia)*

Plate 4.8 Wood-frame construction used for a shopping centre in Yorkshire, England. Douglas fir plywood is used for sheathing and in the construction of plywood web beams. *(Council of Forest Industries of British Columbia)*

Plate 4.9 The lock-keeper's house at Evesham, England, built over a weir. The construction consists of laminated and plywood 'A' frames with stressed skin plywood panels for roof and floor sections. *(Nucleus Projects Ltd.)*

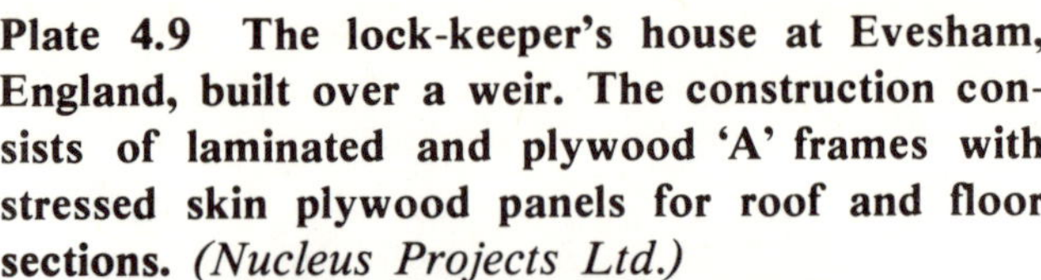

Plate 4.10 Western red cedar shingle roof. *(Council of Forest Industries of British Columbia)*

Plate 4.11 Scots pine, pressure treated with Boliden K33 salt preservative, used in Sweden as external cladding, a relatively cheap and efficient form of construction. *(Anticimexbolagen)*

Plate 4.12 Wood, pressure treated with Boliden K33 salt preservative, used as a durable cladding for a factory building in Sweden. *(Anticimexbolagen)*

Plate 4.13 Wood, pressure treated with Boliden K33 salt preservative, used for the construction of permanent seating and the roof structure in a stadium in Sweden. *(Anticimexbolagen)*

Plate 4.14 Western red cedar used as a wall lining in a bathroom. *(Council of Forest Industries of British Columbia)*

Plate 4.15 West African hardwood plywood used as a decorative wall lining. *(UAC Timber)*

Plate 4.16 Processed plywood factory-finished with a durable weatherproof surface, used as exterior cladding. *(Finnish Plywood Development Association)*

Plate 4.17 Simple laminated beams and pine boarding used in a bar ceiling construction.

Plate 4.18 **A laminated post and beam building under construction.**

Plate 4.19 **Laminated wood staircase at theWood House Museum, Southampton, England.**

Plate 4.20 **Examples of shell roofs.**

as good as the joints between them. Whilst there are many arguments in favour of pre-finished housing units, particularly wall, floor and roof sections which can be rapidly assembled on site, there are only limited arguments in favour of small pre-finished components such as window frames and cladding panels; the protective finish is best applied after installation is complete.

Plate 4.21 A radio mast constructed from laminated wood. This particular mast suffered from fungal decay, clearly indicating the need to preserve all non-durable wood used in exterior situations.

4.3 Wood engineering

Wood has remarkable strength and stability in its longitudinal dimension, and wood engineering aims to take the best advantage of these properties. This strength of solid wood components is seriously affected by defects, particularly the presence of knots, and wood engineering therefore generally involves lamination of thin wood boards so that the influence of these defects can be randomised and reduced to a minimum. Naturally the use of thin boards enables some shaping to be adopted so that large wooden components can be constructed to the dimensions that are required from an engineering point of view. In principle strips of wood are finished to provide a good surface for adhesion and then assembled using cramps to maintain their position while the adhesive cures; chemical curing of the adhesive is generally adopted, but radio frequency systems can be used when rapid curing is required. In the production of beams the wood components are laminated with their grain similarly orientated in order to develop the strength in the longitudinal direction, but in the construction of roof shells a series of planked skins are formed in situ to produce virtually a very large and thick shaped piece of plywood.

In other forms of construction normal plywood is used, particularly as the vertical component or web of composite beams. In the simplest form of beam a single plywood web is joined to top and bottom solid wood components designed to withstand the excessive compression and tension loads at these points. A particularly simple construction is the box beam with plywood glued and nailed on either side of solid upper and lower components, a system that has been adapted to form interesting trough-section roof structures. It is not intended to describe these variations in detail; "Wood Engineering" is the subject of a further book in this series which will comprehensively cover this subject.

Wood based panel products, such as plywood, blockboard, battenboard and laminboard are essentially engineered products in which the natural defects in wood are randomised and the advantageous properties fully developed. However, particle board cannot be considered to be an engineered product; the formation of small particles has necessarily destroyed much of the longitudinal strength in wood components which is perhaps their most advantageous feature.

4.4 Boat building

In recent years there has been a tendency to abandon the use of wood for boat building. This can be largely attributed to the decline in woodworking expertise, for wood is certainly one of the most suitable materials for this purpose. Wood possesses positive buoyancy so that, if a vessel should flood, there is a tendency for it to remain afloat, even in the absence of any buoyancy chambers. Wood is, in addition, very strong in relation to its weight, and its resilience enables it to withstand suddenly applied shock loads; a shock load applied to a steel boat will cause buckling, and glass-reinforced plastic will craze, subsequently absorbing water which often causes severe loss of strength. Wood is easy to machine, shape, bend and finish without the need for special equipment. It gives an excellent appearance and is exceptionally durable when properly used and ventilated. One feature of wood which is not generally appreciated is its excellent resistance to abrasion. Its resilience permits it to absorb some penetration without permanent damage. In addition, abrasion will penetrate into a wood surface at approximately the same rate as into a glass-reinforced plastic surface, yet because of the difference in densities the thickness of a wood-planked hull will be much greater than the thickness of a glass-reinforced plastic hull. The wooden hull is therefore able to withstand severe abrasion for a much longer period than a plastic hull.

Maintenance

Perhaps the greatest contrast between wood and competitive materials arises in connection with maintenance and damage repairs. A major criticism of a wooden boat is the high cost of maintaining paintwork and brightwork (varnish). In fact these difficulties arise solely through the failure of paint manufacturers to produce proper products for finishing wood. They can produce paints and varnishes which give excellent durable glossy surfaces and excellent colour retention but they are largely incapable of producing films which can tolerate the differential movement where an end-grain surface is in contact with a side-grain surface, although in boat-building it is normal practice to employ only wood species possessing small movement in order to avoid these problems. Unfortunately the paint and varnish films are then unable to tolerate preferential wetting. This arises through condensation within the wood or through a minor defect in the film which permits some water penetration. One of the features of wood is that it is hygroscopic, or water absorbing, as explained in Appendix 2. As a result the wood elements prefer to be coated with a polar solvent such as water rather than with a non-polar system such as a paint or varnish coating. As a result any water within the wood tends to spread over the surface of the wood elements, displacing the paint or varnish coating, which then blisters or peels. There are chemical systems which can be used to establish a permanent bond between the paint or varnish film and the wood elements, and if these systems were generally adopted by paint manufacturers this form of failure could be completely avoided. Paint maintenance would then become a simple system of rubbing back the surface and applying a gloss coat in order to preserve the appearance, and this will be precisely the same maintenance that is required for glass-reinforced plastic boats; the gel or finish coat on a plastic boat is particularly susceptible to abrasion damage, and resurfacing with a suitable paint finish will be necessary after several years of use. Indeed, relatively minor abrasion and bruising damage to a glass-reinforced plastic boat will often allow water penetration to occur into the glass fibre reinforcement, perhaps resulting in substantial loss of strength, so that there is a tendency for such plastic boats to deteriorate steadily. In contrast a wooden boat, if properly maintained, will remain in good order continuously, but the greatest contrast arises where damage repair becomes necessary.

Repairs

Where penetration damage has occurred through a hull skin the reinforcement in a glass-reinforced plastic boat has been broken and it is completely impossible for it to be repaired. The normal method is to place a mould over the outside of the boat then to coat it from the inside with a gel coat followed by plastic and glass reinforcement as during the original construction. However, this will not be linked with the reinforcement around the edge of the damage and a patch is therefore added, extending for some distance beyond the damage in order to link the new area with the original surrounding hull skin. There is no continuity in the reinforcement and the repair relies entirely upon the adhesion between the internal patch and the old skin for its integrity. In contrast, the normal method for repairing a wooden boat is to completely remove the damaged components and to replace them with new components, so that the repaired boat is returned to its original condition.

It is true that steel boats can also be repaired by the complete replacement of damaged components but they have other disadvantages in comparison with wood. Naturally corrosion is a very serious problem with steel, which can only be generally avoided by careful maintenance of the paint coating system. Unfortunately minor abrasion damage will allow corrosion to occur and then substantial maintenance is necessary before the affected area can be properly recoated. In addition a steel boat is heavy, noisy and very poorly insulated; even where wood is used for linings there is a danger of condensation on the inner surfaces of the hull and deck which can lead to considerable difficulties.

Wood selection

Whilst it is true that wood is perhaps the ideal material for boat construction it is equally true that little attempt has been made by the trade to develop new and advantageous methods which could enable wood to properly compete with other materials, particularly plastic and steel. There is a considerable danger of fungal decay when wood is used for boat building, not so much in the keel and lower planking of the hull which tends to be permanently waterlogged and thus resistant to decay, but largely in the deck beams where the danger arises through rain water leaks. For this reason boat-building wood should always be durable. In general this means the selection of species possessing natural durability and the careful rejection of all sapwood. Unfortunately some of the traditional durable species are no longer readily available and there have been many instances in recent years of fungal decay in boats through carelessness in selection or the inclusion of sapwood. For example, gedu nohor has often been found in decayed boats in place of the African mahogany that was specified, yet it has very variable durability and is often very susceptible to fungal decay. In a similar manner non-durable maritime pine has been used in place of moderately durable Scots pine. Whilst it must be acknowledged that some changes in boat-building woods are inevitable in view of the supply situation, it is a disgrace that boat-builders completely ignore the possibility of using preserved wood when naturally durable wood is unavailable. Durability is only one of several general requirements for wood used in boat-building; small movement is perhaps the most important additional requirement for any wood used as a board, such as the hull planking and the decking. Finally, it is often necessary to join pieces of wood in boat-building edge to edge and this necessarily means that the wood must have high tensile strength across the grain.

Construction methods

In traditional boat-building the principal components are the keelson and the stem and stern posts. The wood selected for these purposes must be particularly strong and stable across the grain and, for example, oak and afzelia are both used as they possess these properties in addition to excellent durability. The frames which form the cross-sections of the boat are not so critical in properties as they are entirely internal members. They are maintained in position by longitudinal members, generally known as stringers, which are prepared from wood selected because of its ability to bend to the curvature of the boat. This is, of course, one of the requirements of the planking, although steaming is often used as a means to plasticise these components so that they can be bent to the required shape.

Carvel planking

In carvel hull planking and conventional decking the planks fay, or are in contact, edge-to-edge. Generally the edges are arranged so that the joint has a steep V section which can be sealed with a caulking material to make the construction water tight. In some more sophisticated forms of construction a spline is inserted in grooves on the edge of each board in order to span the joint. At the bow the ends of the planks are screwed into a rebate in the stem post and a similar joint is used at the stern in canoe-ended boats; alternatively a flat-board or transom is provided at the stern and the boards are fixed to its edges. In carvel construction every board must be shaped at its edges to permit the planking to follow the contours of the hull.

Clinker planking

In clinker construction the boards are overlapped and clenched; nails are driven through the edges of the overlapping boards and the ends turned to provide a form of riveting. Less shaping is required with clinker planking as the overlap can be varied to permit the planks to follow the hull contours.

Strip planking

These are the conventional forms of construction but there have been several attempts in recent years to utilise wood more efficiently in boat construction. In the strip plank system the width of the planks is reduced, approaching a square section so that it can be bent both in and across its width. The first plank is laid parallel to the edge of the deck and successive planks added, each being bent to take up the required curvature of the hull. The planks do not require to be individually shaped, although it is necessary to chamfer the top edge to ensure a tight joint with the upper strip. Alternatively the strips can be moulded with a concave surface on the upper edge and a convex surface on the lower edge, an arrangement that enables the strips to follow a moderate hull curvature without further shaping and which also gives a large surface for adhesion between neighbouring strips. If these strips are glued and screwed to frames at reasonably close intervals there is no need to have any edge fastenings other than the adhesive, and the hull skin therefore forms a single homogeneous shell of considerable strength. Naturally an appropriate gap-filling adhesive must be employed but this applies to boat-building generally. For example, British Standard Specification 1204: Part 1: 1964 "Synthetic Resin Adhesive; gap-filling phenolic and aminoplastic for constructional work in wood" defines various adhesives and, for boat construction, only the weather- and boil-proof (WBP) types are suitable, although other types could be used for non-structural purposes, such as interior joinery.

Strip planking is, of course, a simple adaptation of conventional planking methods, the main development lying in the use of adhesives to edge joint the boards in place of metal fastenings.

Diagonal planking

In diagonal planking two or more layers of boards are used to form the hull shell, each layer orientated at an angle to the adjacent layer, effectively producing a sheet of plywood out of planks, although the individual planks are trimmed on their edges to enable them to follow the curvature of the hull. Originally a layer of linen coated with oil and white lead was placed between the layers of planks which were then joined to one another with metal fastenings. This form of construction frequently resulted in serious decay between the layers of planks, although this occurred largely in craft constructed in wartime and was the result of the use of unsuitable non-durable woods. In modern construction this problem can be avoided by careful wood selection and, in addition, it is normal to use adhesive between the layers of planks so that the result is a monolithic shell, effectively a tailor-made sheet of plywood. Each layer of planks is generally screwed to the layer beneath in order to ensure a tight and strong adhesive joint.

Cold moulded construction

In cold moulded construction the principle is the same but the planks are reduced in thickness so that they can be more readily moulded to the hull shape. The number of layers are increased to produce the required hull thickness which is then rather stronger than a diagonal planked hull of similar thickness. In practice, diagonal planking is used for larger vessels and cold moulding for smaller, as the thinner planks or veneers in cold moulding more easily follow the sharper curvatures required in yachts and smaller boats. One basic difference between diagonal planking and cold moulding is the lack of fastenings in the latter case; staples or panel-pins are used for temporary fastenings to hold each layer in position whilst the adhesive cures but these are subsequently removed. The process is taken one stage further in hot moulding where even thinner planks or veneers, only 1.5 mm in thickness, are employed in strips or, in effect, very thin planks which are shaped at their edges to follow the curvatures. In this method of manufacture the veneers are laid up, together with the basic longitudinal members such as the edge beams and keels, on a male mould, staples being used to hold each layer in position As a further layer is added the staples are removed so that they only remain in the final outer layer. The mould and prepared shell are covered with a rubber bag in which a vacuum is drawn so that the shell is clamped tightly onto the mould. The bag containing the shell and mould is then placed in an autoclave and heated with steam at high pressure in order to cure the adhesive. This hot moulding technique has the distinct advantage that the first hull shell can be assembled without adhesive and then taken apart in order to provide templates for cutting veneers for future production. It

Hot moulded construction

Plate 4.22 **Wood frames and a plywood skin give an efficient, adaptable and durable system for boat construction, provided proper care is taken in the selection of materials.**

therefore represents the ultimate in efficient wood utilisation and, in effect, a finished hull is a piece of plywood produced to the shape of the hull shell. The technique, first employed for producing aircraft such as the "Mosquito" in World War II, has been used in England for the production of many thousands of craft from dinghies to motor and sailing cruisers. Unfortunately the method is now declining in use, apparently because it is difficult to find staff who are willing to work with wood. This reluctance almost entirely explains the declining use of wood in boat-building, though it is technically efficient for the purpose and, since the petroleum crisis resulted in substantial increases in the cost of plastic, economically attractive as well.

Chine construction

Much of the decline in the use of wood in boat-building can be attributed to the very large use of plywood in the years following World War II. Hulls designed with chines or sharp edges can be constructed from flat sheets of plywood bent in one plane alone. The method of construction is simple, employing relatively light frames to which the plywood is fixed to give the required shape. This method enables boats to be constructed at low cost but relies too heavily upon the edge joints of the plywood. Obviously some of the veneers at the edges are orientated to expose porous end-grain which can rapidly absorb water if improperly sealed. Water absorbed by the veneers will generally cause expansion across the grain and, whilst plywood is designed so that the longitudinal dimension in the adjacent veneer will restrain this movement, the cohesiveness of the wood of the veneers is often inadequate and they break away from the adhesive line. Plywood that has failed in this way is structurally unsound and dangerous as a hull skin. In addition, some of the plywoods used for this purpose have been constructed from non-durable species of wood and the water penetration has therefore resulted in severe fungal decay. These problems can now be almost entirely avoided by the use of improved jointing techniques and by plywood being manufactured to more stringent specifications than in the past; British Standard Specifications 1088 and 4079: 1966 "Plywood for Marine Craft" specify the limited range of tropical hardwoods that are considered to be sufficiently durable or, alternatively, the preservatives that should be employed where non-durable species are used. Plywood remains perhaps the most useful panel material for superstructure construction and it is to be regretted that past carelessness in the plywood manufacturing and boat-building industries resulted in such a severe loss of confidence in wood that glass-reinforced plastic construction became widely adopted. The fact remains that a properly constructed wooden boat will have a life of five or ten times that of a plastic boat, and in the event of serious damage a wooden boat can be restored to its original condition without the use of the doubtful patching techniques which must necessarily be employed for the repair of plastic craft.

4.5 Marine construction

Wood is very widely employed for marine construction, both in erosion control structures and in wharfs, jetties and docks. The principal competitive materials are concrete and steel but, in comparison, wood has remarkable strength-to-weight properties, resilience, abrasion resistance and durability. A wood pile drives well and, in the case of construction on land with limited load bearing properties, its ability to be driven using light hammers with large drops is particularly important as this involves comparatively light pile driving equipment.

River bank erosion control

The banks of rivers in their upper reaches where there is only light traffic can be protected against flooding or scouring due to fast currents by faggotting, bundles of brushwood being installed along the banks by staking or weighting down with stones. This system has been widely used for dyke construction and relies upon the ability of the brushwood faggots to trap silt and eventually form a reasonably permanent bank structure. Fascine mattresses of hazel, willow and birch are also frequently used for riverbank reinforcement. The woods used for these purposes are generally non-durable, although they cannot decay where they remain waterlogged and covered with silt so that they provide an excellent and economic reinforcement for banks. Where traffic on rivers or canals is heavier it is necessary to adopt a far more sophisticated post and horizontal plank construction to form a bank which can resist a heavy wash. There is a severe decay risk where the wood is above the water level and it is essential that naturally durable or adequately preserved wood should always be used. In the case of estuarine conditions there is, in addition, a risk of marine borer attack in the wood below the water level and in this case it is often more realistic to use a species that can be readily impregnated with a preservative as it is comparatively difficult to find woods that have natural resistance to marine borers, particularly gribble, *Limnoria* spp.

Coastal erosion control

Structures to prevent coastal erosion are more sophisticated because of the much greater forces that they must be able to resist. It is not sufficient for a sea defence such as a groyne to survive as it must also be designed to oppose the drift of material parallel to the beach, as it is these lateral currents that account for most of the beach erosion. There is also the direct frontal effect of the waves on the beach, tending to accumulate sand in the summer months then remove it in subsequent winter storms. A groyne consists essentially of a wall running down the beach, about 100 to 300 ft. (30 to 90 m.) in length with about the same distances between each groyne. The height naturally depends upon the local conditions but is typically about 3 ft. (1 m.). A simple groyne is normally constructed using 9 in. (225 mm.) square piles, 10 to 20 ft. (3 to 6 m.) long and spaced 6 to 9 ft. (2 to 3 m.) apart. The spaces between these piles are then filled with horizontal planking, typically 6 in. (150 mm.) wide and about 2½ in. (62.5 mm.) thick. Further horizontal waling members are then attached to the base of the groyne or perhaps higher up, depending upon the individual design. In some cases land ties are provided, consisting of props on the sheltered side designed to prevent the groyne from being flattened by an accumulation of beach material on the weather side; the end of each prop is typically supported by two posts driven into the beach. Sometimes permeable groynes are used, consisting of vertical poles supported between horizontal walings. In some cases a zigzag form of groyne is preferred to a normal straight wall. In addition wave screens are sometimes erected parallel to the shore in order to resist the drawdown in stormy weather, which tends to remove material from the beach, but they are rather unpopular as they tend to interfere with the amenity value of a beach. Finally revetments and breastworks are often constructed parallel to the shore but above high water mark. These normally consist of a pair of impermeable walls constructed from piles, waling and planking. The space between the walls is then filled with earth, stone or simply beach material so that this forms a massive defence against severe storm penetration of the beach.

Plate 4.23 Groynes in Wales constructed from opepe, a West African hardwood. *(UAC Timber)*

Plate 4.24 Pine, pressure treated with Boliden K33 salt preservative, used for piles, beams and decking of a light-traffic yacht jetty. *(Anticimexbolagen)*

Wharf and jetty construction

Wharf and jetty construction involves the use of piles, fenders and decking as well as beams and braces. Piles are used as the load bearing supports for jetties, as moorings and as fenders to protect wharfs and jetties. Generally very large diameter wood is used which can be expensive because of its limited availability, although serious consideration is now being given to the use of laminated wood for this purpose. In the British Isles, for example, it is traditional to use square piles, prepared either by sawing or hewing, whereas in North America it is traditional to use round piles. Both systems have their advantages; round piles are less expensive to prepare but more difficult to assemble. In round piles it is essential to pressure impregnate with a suitable preservative as the outer sapwood layer is invariably non-durable. However, it is almost impossible to obtain wood that is naturally resistant to marine borer attack so that preservation may be essential in any case, and the sapwood remaining on a round pile will generally be more efficiently preserved than heartwood exposed in a squared pile.

Fenders take several forms and are invariably constructed of wood which is the only structural material which possesses the necessary resilience and abrasion resistance to withstand and absorb the shock of collision damage. Fender piles are often provided around a jetty in order to protect the load bearing piles. Vertical fenders are often attached to the concrete face of a quay, or horizontal wood fenders are provided on concrete and steel piles, really acting as rubbing pieces, so they require considerable resistance to splitting. In many cases the fenders take the form of floating booms attached to the structural piles but in all cases it is the ability to absorb and withstand collision damage that ensures that wood is invariably used for these purposes. The choice of wood is, however, limited. Non-durable wood should never be used as there is a danger of splitting which may expose untreated inner areas. In addition, woods with interlocking grain, such as afzelia, and wavy grain, such as elm, are usually used as they tend to possess good resistance to abrasion.

Braces and beams do not require any special mention as they represent a normal structural use for wood, but there are decking problems. A continuous impermeable decking will tend to remain very wet with, as a consequence, a high decay risk. This risk can be reduced by providing proper drainage, normally by separating individual planks by gaps, but natural durability or adequate preservation is essential. In addition the decking must have adequate resistance to wear and strength to support heavy loads; jetties and wharfs must frequently carry heavy wheeled traffic, both when working ships and during construction and maintenance work. In recent years one of the largest uses for decking has been in the many marinas that have been constructed to provide moorings for pleasure yachts. Naturally the requirements for marina decking are very similar to those for wharf and jetty decking, except that the loads are likely to be much lighter, and there is a particular requirement that the surface should not become slippery. This presents considerable difficulty as the choice of a species which can be sawn to give a reasonably smooth but skid-resistant surface is often insufficient on its own; marina decks in mild wet climates often become colonized by algae which must be controlled chemically.

Marina construction

The choice of species for marina work depends upon their function. For example, piles must be constructed from wood which is readily available in long, straight-grained pieces of large cross-section with excellent compressive strength parallel to the grain, such as greenheart. In contrast fenders are not necessarily required in long lengths but they must possess wavy or interlocking grain to give good abrasion resistance. European oak is often used for this purpose or alternatively elm which is very similar in properties but rather less strong and much less durable. Decking must be made from wood which is easy to saw and which gives a reasonably smooth surface, free from splits and splinters, such as, for example, keruing which also possesses good resistance to even heavy pedestrian traffic.

4.6 Railway and road construction

All wood in railway and road construction work must be naturally durable or adequately treated with a suitable preservative. Provided this basic precaution is observed wood is efficient and economic for these heavy construction purposes. Retaining walls and revetments, similar to those previously described for coastal or riverbank work, are widely used throughout the world, as wood has very considerable advantages over competitive concrete and steel structures. These advantages are not solely in terms of cost but also in the technical advantages of wood, particularly its elasticity and resilience, combined with the ease with which it can be worked on site compared with prefabricated concrete or steel members. In addition its simple jointing and comparatively low density mean that relatively small and light components can be transported, perhaps across very difficult terrain.

Bridge construction

Bridges provide an excellent example of the use of wood as an engineering material. The simplest bridge consists of two beams with a decking fixed to their upper surface, but as spans increase bridges must be engineered in a more critical manner. One method is to provide pile supports at regular intervals, using beams to span the spaces between them. In many instances simple piles are unsuitable and substantial trestles must be constructed, the bridge itself being trussed in a complex manner to enable it to span a wide gap. Alternatively large laminated arches may be provided beneath the bridge as the principal supports for the posts carrying the deck beams above, thus providing a clear span, but non-trussed construction. The assembly of bridges is relatively simple for most designs, as joints are only necessary to keep the

Plate 4.25 **A footbridge supported by laminated wood beams.**

posts and beams in position; the actual load is taken through the contacting surfaces between the posts and beams. However, in the case of truss construction, more complex joints are necessary to enable some of the members to be in tension. In modern construction this generally means the use of steel plates and multiple bolts in order to provide tension joints between one component and another.

The woods used for the various components in bridge construction are selected according to their properties in the normal way. For example, the piles must be straight grained and possess good compressive strength parallel to the grain. The beams must have good strength in bending whilst some of the truss components must have good strength in tension. In contrast the decking must have good wear resistance, and it may, in fact, be short lengths of wood with very distorted grain that will prove most suitable for this purpose. In all cases the wood must be naturally durable or readily preserved, and this certainly represents one of the major requirements when selecting wood for construction purposes of this type where there is a severe decay hazard.

Railway sleepers (ties)

Wood is particularly suitable for use as sleepers (ties) in railway construction. Sleepers are subjected to very substantial impact loads in contact with the flat bed of the rail or, where they are used, the chairs which support the rails, so that the resilience and elasticity of wood is a considerable advantage. In addition the rails or their chairs must be fixed to the sleepers using spikes or screws and wood has the advantage of very good holding, which simplifies this fixing problem. Various species are commonly employed, such as beech by the German railways, Scots pine by the British Railways, Douglas fir and hemlock in North America and various Eucalypt species in Australia. Each of these species presents a different woodworking and preservation problem, but the greatest difficulty is the immense volume of wood that is required. Admittedly efficient preservation reduces the demand for replacement wood but it must be appreciated that fungal decay is not the only destructive factor; violent fluctuations in conditions between hot, dry weather and heavy rainfall may result in severe mechanical damage through movement, and the pounding under the rail may combine with this movement damage to cause mechanical failure. For this reason salt preservatives are unpopular for sleeper preservation as they provide only protection against fungal decay, whereas creosote gives good control of movement. In recent years the increasing cost and declining availability of wood of adequate dimensions, coupled with creosote shortages, has led to a decreased utilisation of wood throughout the world, despite the efficiency of creosote treatment in reducing demand for replacement sleepers. Some steel sleepers have been used, but concrete has become very popular and has now been widely used for many years. It is far less efficient than creosote-treated wood and is in fact completely unsuitable for use in new high speed tracks and, as a result, there is now a tendency to return to the widespread use of wooden sleepers.

Road paving blocks

Wood has been used in the past as a road surfacing material. Generally wood blocks, heavily impregnated with creosote, give an excellent wearing and resilient surface when laid with an end-grain surface upwards. Unfortunately the laying of wood blocks involves a very high labour element which can now no longer be afforded and wood blocks are now only used in heavy industrial premises where their exceptional wear resistance and resilience is of particular importance.

Fences

Road furniture represents a large volume use for wood. The obvious use is in fences where durable or preserved wood is widely employed. Concrete is the principal competitor, yet it certainly has no greater life than naturally durable or properly preserved wood and it has the distinct disadvantage of the weight of the posts and other components. Indeed, it is weight that prevents concrete being widely used as a replacement for boarded fences. Fencing techniques naturally vary throughout the world, but wood is suitable, whatever the local conditions. Round poles with wire can provide the simplest fence, but a few years ago cleft wood, particularly chestnut, was almost invariably used for posts, rails and boarding for fences in Britain. Sawn wood is now normally used, but naturally durable home grown oak is still preferred in Britain,whereas preservative-treated softwood is preferred, for example, in Scandinavia. Other uses of wood are for the small triangular posts fitted with reflector plates, that are frequently installed alongside fast roads in order to discourage deer and other wild animals from crossing; the headlights of approaching traffic are reflected by the plates fixed to the posts so that the wild animals are discouraged from approaching too close to the road. Another use for wood is in crash barriers where wooden posts are able to absorb impact shock to a much greater extent than either steel or concrete can. The shock-absorbing capacity of wooden crash rails is also an advantage, although in most countries metal crash rails are used fixed to wooden posts.

4.7 Poles, posts and piles

In many respects poles, posts and piles represent similar technical problems as they all involve a ground-line condition where there is a very severe fungal decay danger. In other respects they only vary in their dimensions and in the fact that poles have most of their length above the ground in contrast with piles which usually have most of their length below the ground.

Fence posts

Fences often receive very little attention, yet they represent a very large volume use for wood. Usually local species are employed, and there is thus a tendency for their value to be largely ignored as replacements can be readily obtained. In fact, any replacement or repair of a fence represents substantial labour costs and there is always justification for the selection of naturally durable wood or the use of a preservation process. This comment applies particularly to the posts as the greatest fungal decay risk is at the ground-line. Rails and boards are less likely to decay, except where rails are mortised into posts and there is a danger of moisture being trapped in the joint.

Transmission poles

Round wood transmission poles are used throughout the world. The principal advantages of wood are its excellent strength-to-weight properties and its elasticity under load. Naturally durable wood is used comparatively rarely and most poles are vacuum/pressure treated with creosote or water borne salt preservatives. Creosote has the distinct disadvantage that it is dirty and likely to bleed, perhaps causing serious damage to clothing where poles are erected in areas with heavy pedestrian traffic. The bleeding is reduced if the Rüping or Lowry empty-cell processes are used, although the pole can never be completely clean. The water borne salts give clean poles but little protection against movement through changes in moisture content. As a result, salt-treated poles tend to split and there is a danger that these splits will penetrate through the treated sapwood into the untreated heartwood beneath. In Scots pine poles, for example, the heartwood is moderately durable but still susceptible to slow fungal decay close to the ground-line under these conditions. In some other species, such as spruce, it is difficult to obtain penetration even into the sapwood, yet these splits may penetrate into the heartwood which is also non-durable, and spruce can therefore be considered to be entirely unsuitable for use as transmission poles. It is true that incising the outer surface of the pole will improve the penetration but this is only a suitable method in the case of, for example, Douglas fir (Colombian pine) which possesses heartwood with very good durability.

Cross-arms are generally produced from solid sawn wood, either naturally durable or pressure treated with preservative. In theory poles should never be trimmed on site so that there is no danger of working exposing untreated wood. In practice, some trimming of the sapwood invariably occurs to prepare a bed for the cross-arms and a hole must be drilled through the pole so that they can be secured. In addition poles sometimes fail because they have been trimmed to length either at the top or the bottom, although there is no excuse for this trimming as poles of various sizes are normally available from stock. In the case of high tension transmission lines, two poles may be used, fitted with substantial cross-arms in order to support the heavy insulators and wiring. At the present time cross-arms are solid wood or steel but it has been suggested that laminated wood might be more suitable for this purpose. In some rocky areas it is impossible to provide a deep hole for the post and bracing wires must be used or, alternatively, foundation baulks; these baulks are secured with bolts and bracing pieces to the post below ground level and are provided with blocks at either end in order to stabilise the post. One of the advantages of wood lies in its insulation properties, particularly when treated with creosote rather than water borne salts, but in high tension cables it is necessary to ensure that the wire stays are also insulated. A laminated wood beam, impregnated with resin, is therefore incorporated within the stay as an insulator, with a spark gap to provide a discharge route should lightning strike the cable. Finally cattle guards round stays and posts are often constructed from wood, so that it will be appreciated that wood has a large number of applications in connection with power distribution and telegraphic communications. It is perhaps worth adding that another substantial use for wood in these industries is in the manufacture of cable drums, usually from softwoods impregnated with water borne salt preservatives.

Piles

Piles have been mentioned in previous sections and need not be covered in detail here. However, it is perhaps interesting to note that steel tubular piles and concrete piles are widely used in Europe, yet pressure creosoted round wood piles are preferred in North America. It is possible that this is related in part to the availability and cost of long piles in each of these areas. Alternatively it is possible that the failure to use proper preservation techniques has resulted in wood having a poor reputation in Europe. In contrast the use of pressure impregnated round wood piles is a comparatively recent introduction in North America and has been successful because of the universal use of proper preservation techniques.

4.8 Shuttering, shoring and scaffolding

Shoring

Previous sections of this chapter have considered the use of wood for structural purposes but they have omitted to comment upon the use of wood as a temporary support during construction. The first use is in shoring, or the construction of load bearing linings to trenches and other excavations in order to prevent their collapse during working. Wood is particularly suitable for this purpose as it is light, strong and very easily cut on site to the required dimensions.. Generally shoring consists of horizontal or vertical boards, supported by posts or rails respectively, with struts across the excavation to maintain the separation. The principal load concentration is on the struts, a function for which wood is particularly suitable in view of its very high strength in compression along the grain. In addition wood has particular advantages as the planking, as any excessive loading will generate splintering and cracking noises which will give warning of an imminent collapse. This is

Plate 4.26

Extensive use of plywood shuttering and pine scaffold boards in the construction of a reinforced concrete nuclear building. *(Nuclear Power Group and Finnish Plywood Development Association)*

considered to be one of the principal advantages of wood in mining operations; a wooden pit prop will always give a loud warning in advance of collapse whereas steel will simply bend and collapse instantaneously when under an excessive load.

Shuttering

Steel shuttering is sometimes used for the construction of form work for casting concrete in situ but it is only economic if it can be re-used, and it is thus only suitable for repetitive work. In contrast wooden form-work can be constructed on site to the shape and dimensions required. One problem with wood is the absorption of water from the cast concrete which causes the wood to swell but also removes water from the concrete surface, perhaps causing serious weakening. In addition the concrete may establish a substantial bond to the wood but all these problems can be avoided provided the wood surface in contact with the concrete is smooth and properly sealed with a shuttering oil or release agent. Occasionally coarse-grained wood is used deliberately to transfer a wood grain effect to the concrete surface, but generally a smooth finish is required. Plywood is now very widely used for the construction of form work, and special film-faced plywoods are now available, surfaced with a phenol-formaldehyde film which ensures good concrete release.

Hoarding and scaffolding

Very large quantities of wood are also used in scaffolding, particularly for scaffold boards. Wood posts, rails and braces are also used in many countries, usually where wood is readily available locally whilst steel poles represent an expensive import. Plywood is also used as hoarding or fencing around construction sites.

Plate 4.27

Plywood used for a hoarding or protection around a construction site. *(Finnish Plywood Development Association)*

4.9 Packaging

Wood is used so universally for packaging that it hardly requires any detailed mention in a book of this type. The main advantage of wood is its excellent strength-to-weight ratio, coupled with the ease of woodworking; it can be readily assembled into special crates with individual requirements or, alternatively, standard mass-produced crates can be simply and economically manufactured. Perhaps the traditional tea-chest provides the extreme example of the well-established utilisation of wood in a particularly efficient manner; it represents one of the first large-scale uses for plywood. In recent years large sheets of plywood and solid wood have both been employed in the manufacture of very big crates for transporting vehicle components, yet the smallest and most valuable items are also protected in wooden packages, perhaps machined into compartments to contain small individual items such as feeler gauges.

Naturally large crates are expensive and in recent years there has been an increasing tendency for them to be re-used. This has resulted in the development of containerisation, in which permanent crates or containers can be delivered door-to-door, and then re-used for a further delivery. Some of these containers are constructed from plywood, the roof-sections perhaps being specially manufactured with the required curve. In other forms of construction aluminium is widely used but, in almost all cases, the flooring sections are constructed from wood. The use of wood in containers presents certain problems when they are likely to be used on routes to Australia; the Australian quarantine requirements have been explained in more detail in Chapter 3.

4.10 Agriculture

Agricultural buildings have long been traditionally constructed from wood in all parts of the world, varying from the log buildings constructed in the mountainous regions of the Alps and Scandinavia to the simple frame and clapboard-cladded buildings used almost universally. In recent years there

Plate 4.28 Plywood used for the construction of an insulated container. *(Finnish Plywood Development Association)*

Plate 4.29 Poles, pressure treated with Boliden K33 salt preservative, used for the construction of agricultural buildings in Sweden. *(Anticimex-bolagen)*

Pole buildings

has been a tendency to abandon the use of frame buildings and to adopt instead a pole construction, typified by the Dutch barn consisting of simply a roof supported on a number of poles, perhaps with one or more wall panels. When the Dutch barn was first introduced it was constructed from steel but concrete has also been used as well as laminated wood. However, the simplest and most economic form of construction is certainly the pole barn in which pressure-treated round wood is used for the basic structural members. The use of preserved clapboard cladding enables this form of construction to be extended to other purposes, such as cow-houses and implement sheds, although plywood is also used as a simple wall cladding.

Plate 4.30 Farm buildings clad in preserved birch plywood. *(Finnish Plywood Development Association)*

Plate 4.31 Silos constructed in Sweden from Scots pine pressure treated with Boliden K33 salt preservative. *(Anticimexbolagen)*

There are a multitude of small uses for wood in agriculture. Plywood is particularly suitable for the construction of curved silos, or for wall linings in silage pits. However, one of the major uses for wood in agriculture is certainly in fencing, but this is a subject that has been considered in a previous section. Gates must also be mentioned, and the need to use naturally durable or reliably preserved wood; a well ventilated gate would appear to present a relatively light fungal decay risk, yet the mortice and tenon joints in the gate will trap moisture and fungal decay, and collapse becomes inevitable if non-durable wood is employed.

4.11 Vehicles

Commercial vehicles

Wood is the traditional material for the construction of vehicles of all types. In recent years car (automobile) body construction in wood has been abandoned in favour of pressed steel and welded construction but wood is now very widely used for commercial vehicle construction. The main advantage of wood is its excellent strength-to-weight ratio and its rigidity which makes it particularly suitable for floor and wagon side construction, even when other structural components are metal. Rail wagons are a particular example; pressed, riveted and welded sides have been employed but wood remains the preferred material for the floors. Solid wood is perhaps most widely used, though the floor members of rail and road wagons, and containers, are now often laminated. Alternatively heavy plywood can be employed, perhaps finished with a special surface such as a moulded rubber. Processed plywoods are also used for body side construction; film-faced plywood is often used so that the face provides an inner surface and only the outer surface requires painting. Curved plywood roof-sections are also available.

Caravans (trailers) represent a rather special form of construction as the lightest possible weight is essential. Thin plywood or hardboard is frequently used on light wood frames such as spruce. Wood is also almost universally used for the interior joinery or furnishings in caravans and it is almost impossible to envisage any alternative material that might be employed. Glass-reinforced plastic has been used for both caravan and light van construction but it generally has little advantage over wood; if a long production run is required metal pressing is undoubtedly most economic but with relatively short runs wood and plywood are most suitable.

4.12 Furniture

Anyone who has sat on a low cost metal or plastic chair will appreciate the principal advantages of wood in furnishing. Firstly, wood used in the same dimensions as metal or plastic will give a far lighter weight or, if it is used at a similar weight, it will achieve greater rigidity. Secondly, the low conductivity and low specific heat per unit volume ensure that wood is warm to the touch as well as being decorative. There have been many developments in recent years which have improved the utilisation of wood in furnishing. The principal development has been the almost complete abandoning of solid wood panelling in favour of plywood or chipboard faced with decorative veneers or, in the case of functional furnishings, with hard wearing plastic laminates. In addition laminated components have been introduced, enabling the conventional furniture construction with its dependence upon joints to be largely abandoned for some purposes. For example, a front and rear chair leg with a joining beam to support the chair seat might be constructed from a single laminated member moulded to the required shape.

Plate 4.32 **Modern furniture simply constructed using birch plywood.** *(Finnish Plywood Development Association)*

5. Commercial Woods and Wood Products

5.1 Typical wood properties

Woods are divided commercially into the softwoods or conifers and the hardwoods or dicotyledons. In addition it is possible to apply a further geographical division. Thus the pines are found wherever suitable conditions exist but the Southern Hemisphere species are generally different from those in the Northern Hemisphere, although the situation is now somewhat confused due to the introduction of species from one country to another. In the case of hardwoods, a wood that is commercially described for example as an oak may belong to one family in one hemisphere and a totally different family in the other, as in the case of the oaks and beeches available in Europe and Australia respectively. However, there is a logical reason for applying the same commercial name to these entirely different woods as, despite their different botanical classification, they are usually superficially similar in respect to their appearance and mechanical properties.

Scots pine

Most of the commercial woods of the world are listed in Appendix 3, together with a summary of their principal properties and sources. The purpose of this chapter is therefore to describe some of the more important commerical woods in greater detail. The softwoods are the principal construction woods of the world, generally possessing exceptional strength in bending so that they are particularly suitable for use as beams. Scots pine or Baltic redwood, *Pinus sylvestris,* is the softwood that is most widely used in Europe. It is used for almost all purposes in building construction, including beams, joists, rafters, internal and external joinery (millwork), and floorboards. In addition it is used for railway sleepers (ties), pit props, general ship-building purposes and transmission poles. In fact, it is used for almost all non-decorative purposes, although it is occasionally selected as a decorative wood, generally when it is rather knotty. It has moderate weight with pale reddish or brownish heartwood and lighter coloured sapwood. It is resinous, moderately hard, fairly strong and seasons well. The heartwood is moderately durable and the comparatively wide sapwood readily absorbs preservatives, so that this is an ideal wood for use where durability is essential as it can be readily achieved by simply treating the sapwood, as in transmission poles. Scots pine is easy to work and paints well. It is largely obtained from the Baltic area and Russia but the close-ringed wood from the extreme North, such as the Kara Sea, has lower strength, although its fine texture and excellent workability make it particularly suitable for joinery (millwork). Scots pine from more southerly areas is coarser in texture but stronger, although wood from the extreme south of the growing area such as England is rather coarse in texture with a somewhat higher movement than the northern wood.

Spruce

Scots pine is becoming scarce and there are serious attempts to substitute it by spruce, white wood or white deal, *Picea* spp., both because spruce is available from the same forests as Scots pine and also because it grows more rapidly in plantations. The principal spruce from Northern Europe is Norway spruce, *Picea alba,* but Sitka spruce, *P. Sitchensis* is widely grown in North America and also in the British Isles. Spruce is white, pale yellow or pinkish white in colour, even-grained and easy to work, except where small, hard, dark knots occur; for this reason it is particularly important to trim off the lower branches when growing in plantations. Unfortunately spruce is non-durable, yet both heartwood and sapwood are relatively impermeable and cannot be readily treated with preservatives, so that this species is largely unsuitable for any purpose in which durability is required. This is especially unfortunate as spruce is, in other respects, ideal for use as transmission poles. Spruce is therefore used largely for joinery (millwork), particularly for interior work, for general carcassing (building framing) and for packing cases; it is almost free from odour and useful for packaging for food stuffs. The straight grain, relatively low density and high strength of spruce make it particularly suitable for spars. When it is finished it has a more lustrous appearance than redwood.

Corsican pine

Many other pines are being used as a substitute for Scots pine. For example, Corsican pine, *Pinus nigra,* is used for many purposes but it has a large proportion of sapwood so that it is generally used for non-critical purposes, such as interior joinery (millwork). Various firs, *Abies* spp., are also being used to an increasing extent as a Scots pine substitute. In many respects they are similar to spruce as they are relatively light, smooth working, yellowish white, non-resinous and free from odour. These comments are, however, largely confined to the European situation, although the importance of Scots pine extends far beyond Europe as it is traditionally exported throughout the world as a construction wood, although exports beyond Europe are now largely confined to preserved transmission poles. The other major sources of softwood have always been Canada and the United States of America where the most important wood is probably Douglas fir, Oregon pine or Columbian pine, *Pseudotsuga menziesii,* from the west coast. Douglas fir grows as large trees so that this pinkish wood is available in very large dimensions, including great lengths; a flagstaff at Kew in England is 200 ft. high. The heartwood is very durable and the relatively narrow sapwood can be treated by pressure impregnation after incising. One feature of this wood is the irregular cambium; examination of a cross-section will show the wavy nature of the annual rings. Douglas fir is widely used for the production of plywood and, as a result of these irregular growth rings, peeled veneers often show blister grain. Generally Douglas fir is used for the same purposes as Scots pine, for example general construction and joinery (millwork), transmission poles and railway sleepers (ties), both in North America and in other countries to which it is exported. In view of the straight grain of the heartwood it is particularly suitable for use in bridge construction and in ladder sides.

European firs

Douglas fir

Southern pine

Whilst Douglas fir is used widely throughout North America as the preferred structural softwood, a large number of other species is also used. Southern pine includes a very large number of species growing in the southern United States, including long-leaf, short-leaf, loblolly, slash, pitch and pond pines; these various woods are identified in the general index and Appendix 3. The heartwood colours of these various species range through yellow and orange to reddish brown or light brown. Most of these species are resinous but fairly strong so that they are ideal for general constructional purposes, as well as railway sleepers (ties) and piles if properly preserved. Many other pines are widely used for general structural purposes in North America, such as the sugar, western white, eastern white, lodgepole, ponderosa, red and jack pines. However, one additional softwood that should be mentioned in detail is hemlock. This occurs as eastern or Canadian hemlock, *Tsuga canadensis,* and western hemlock, *Tsuga heterophylla,* both of which are used for general constructional purposes. Where greater durability is required Californian redwood, *Sequoia* spp., may be selected with its dark reddish colour, although it is relatively soft and rather lower in strength than, for example, Douglas fir.

Hemlock

Rimu

Kauri

Radiata pine

Patula pine

In the southern hemisphere rimu, or red pine, *Dacrydium cupressinum,* is the principal indigenous softwood of New Zealand, possessing excellent properties very similar to those of Scots pine. Unfortunately the growth rate is rather slow and felling has exceeded natural renewals for many generations. Kauri, *Agathis* spp., are also important softwoods in Australasia, with a finer texture and greater stability then rimu. Kauri is becoming increasingly scarce and is now used largely for the production of relatively high value fittings and ship decking. With the pressures on local sources it is not surprising to find that Australasia is now having to import, for example, kauri from Indonesia and there have been considerable efforts in recent years to introduce more rapidly growing softwood species to meet the demand for wood for general structural purposes. The most important of these introduced species is radiata, insignis or Monterey pine, *Pinus radiata,* which can now be found in plantations in many parts of Australia. The wood is white to creamy yellow and, in addition to general structural uses, it possesses the distinct advantage that it can be readily preserved and then used in situations where there is a severe risk of fungal decay. One problem with radiata pine is its comparatively high movement. Patula pine, *Pinus patula,* is another fast growing pine introduced into Australia, but it is most widely grown in South Africa where it is known as South African or S.A. pine. It grows extremely rapidly, giving very wide growth rings and generally achieving an economic diameter for felling before significant amounts of heartwood have been formed. As a result this wood is non-durable, yet it is very readily preserved and finds widespread use as a construction wood in South Africa in areas where the local regulations require preservative treatment to be applied as there is a severe risk of termite or Longhorn beetle attack.

Larch

American cypress

There are three softwoods that should be mentioned as they are often selected for use because of their natural durability. Larch, *Larix* spp., are relatively heavy softwoods with heartwood darker in colour and rather more durable than that of Scots pine and a thin layer of sapwood that can be readily treated with preservative after incising. The wood is fairly soft and easy to work, yet tough so that it finds applications as poles or piling, particularly in sea and river erosion control. Siberian larch, principally heartwood, has a very fine grain which, combined with its natural durability, makes this wood particularly suitable for the manufacture of exterior joinery (millwork). Another wood that finds very similar uses is American cypress, *Taxodium* spp., with a yellowish brown heartwood which is also very durable. However, the best known wood for natural durability is perhaps western red cedar, *Thuja plicata,* from British Columbia. This is a very light wood, easily worked and with excellent natural durability which causes it to be exported throughout the world. Usually these exports are in the form of planks, roofing shingles and shakes, or milled woodwork. Western red cedar is widely used for light construction purposes and also for external cladding where it is frequently weathered naturally; it eventually reaches a light silver-grey appearance, but protective stains are sometimes used to retain the red colouration. The wood has relatively low movement and a pleasant odour which makes it particularly attractive for use as a panelling material, although one problem with western red cedar is danger of collapse during seasoning.

Western red cedar

Oak

Amongst the hardwoods oak, *Quercus* spp. are perhaps most widely known. In most structural uses chestnut, *Castanea* spp. are a traditional alternative; the appearance and properties are very similar to oak but chestnut can generally be readily distinguished through the absence of medullary rays or silver figure. Oak is used for both heavy and light structural purposes, floor boards and blocks, railway wagons and truck bodies, boat frames and planking, marine groynes and piles, bridges, fences and gates, barrels and coffins. The bark is used in addition as a source of tannin and, in the case of cork oak, the use of the bark is obvious. English oak tends to be a shorter and broader tree than the other oaks used for commercial wood, giving a wood that has wide rings but which is strong and durable. English oak is normally a medium brown colour with a relatively coarse texture but with a particularly fine silver figure on the radial or quarter-sawn faces which makes it particularly suitable for oak panelling; quarter-sawn veneers are used in decorative furnishings. Austrian oak is a uniform golden brown colour with rather straighter grain than English oak which makes it particularly attractive for high quality joinery (millwork) and panelling. In fact, the use of Austrian oak for panelling has resulted in its normally being delivered as wainscot billets, the most economic way for developing the attractive silver figure. Various oaks are also available from America, varying from hard, coarse but strong wood which is used for the manufacture of furniture and joinery (millwork), to slow-grown even-textured wood which is used particularly for strip floors. Generally American oak is classified as white oak, pale brown in colour with the vessels blocked by tyloses giving a hard impermeable wood, and red oak, pinkish in colour and lacking tyloses so that it is relatively soft and so permeable that smoke can be blown through the vessels. Japanese oak is slow-grown, uniform in colour and easily worked so that it is particularly suitable for high quality furniture and joinery (millwork). In contrast, Australian silky or Tasmanian oak is a Eucalypt species which is relatively soft and coarse in texture, its only similarity to true oak being its silver figure which makes it valuable for use in furniture, both as solid wood and veneers.

Australian oak

Ash

Ash, *Fraxinus* spp. give fairly heavy, straight-grained woods which are usually pinkish when cut, although becoming yellow or pale brown. The wood is hard, tough, elastic, durable and hard wearing. It is particularly suitable for the manufacture of tool handles because of its elasticity and excellent resistance to shock. Ash is grown throughout Europe, Canada, America and Japan, although in American hickory, *Carya* spp. are very similar in properties and find similar uses. Tasmanian or mountain ash, the

Hickory

Australian equivalent, is unrelated as it is a Eucalypt species with only superficial similarities, although again used for similar purposes.

Elm

Elm, *Ulmus* spp. generally give warm brownish woods which are moderately dense, strong and with irregular grain which tends to resist splitting when subjected to impact or abrasion damage. Elm is non-durable in moist conditions, yet it is exceptionally durable when waterlogged, and finds many applications in dock and wharf construction. It is also used as weatherboard cladding, as well as for the production of coffins, chair seats and chair backs. Wych Elm and Canadian rock elm are more durable and are both used for boat frames and planks. Dutch elm is similar in basic properties but rather paler in colour.

Beech

Australian beech

Beech, *Fagus sylvatica,* is a strong, hard and reddish or pale pinkish wood, darker in colour when steamed. It is a close-textured, even-grained wood with a small silver figure apparent on radial or quarter-sawn surfaces. Beech suffers from a rather high tangential shrinkage and distorts badly when flat sawn pieces are dried. It will not withstand fluctuations in moisture content and is non-durable so that it is generally employed for small technical or domestic objects, such as tools and furniture, particularly chair legs. Beech is produced in both Europe and North America, and a related species, *Fagus crenata,* grows in Japan. Tasmanian, Southland and Antartic beech are *Nothofagus* species in the Southern Hemisphere which are superficially similar to beech in appearance; they are more correctly described as myrtles and differ from true beech in the absence of rays.

Birch

Willow

Lime

American basswood

Maple

American whitewood

Birch, *Betula* spp. are fairly light and yellowish or reddish in colour. Birch is generally used for furniture or small articles in which the fine grain is an advantage. In addition birch is widely used in the Baltic countries for the production of plywood and, in Canada and the United States, solid birch is used as a general utility timber, particularly for furniture. Birch is non-durable and tends to chip and splinter relatively easily so that it is most suitable for uses where it will be finished with paint or varnish. Willow, *Salix* spp. are soft and similar in appearance to birch, but have a fundamental difference in their excellent resistance to splitting and splintering on impact, so that willow is used for the production of cricket bats and for lining wagons or trucks. Willow was also at one time widely used for brake blocks, particularly on the railways, as it does not polish when rubbed by the steel wheels and it does not burst into flames when over-heated. Lime, *Tilia* spp. are also soft, light-coloured, close-grained and very even textured; they are known as basswood in America where they are used for the manufacture of plywood, producing a product similar to the European birch plywood, and also for other manufacturing purposes where the fine uniform texture is an advantage. Maple, *Acer* spp. are also similarly used in Europe and North America, perhaps surprisingly as a decorative veneer, although only when a very light colour and uniform appearance is desirable. However, the most popular soft, even-grained and easily worked wood in North America is whitewood or tulip tree, *Liriodendron tulipifera*. This wood is variously known as American or Canary whitewood, or yellow poplar; it has a creamy colour when cut but becomes yellow to brown when exposed to the air. It is widely used for joinery (millwork) and furniture manufacture as it surfaces particularly well.

Walnut

Walnut, *Juglans* spp. are particulary important in furniture manufacture. They are generally heavy dark brown woods, although the lighter-coloured sapwood is, like oak, susceptible to Powder Post beetle attack. Walnuts are strong, hard and tough with a very fine texture. They season well and are very stable, so that they are suitable for the production of rifle stocks and furniture, particularly as they are easy to work and take a high polish. Walnuts are also popular for the production of solid or veneered panels. The normal sources are Europe and North America, the various woods often being described in terms of their figure, such as mottled, crotch, curl and feather, although burr or birds eye, generally Caucasian walnut, is perhaps most highly prized as a furniture veneer. American black walnut has particularly straight grain with a good colour and excellent stability which makes it widely used for domestic furniture and shopfittings.

Australian walnut

Typical walnuts

There are a large number of woods which are described as walnut, although they are not *Juglans* species. These include the satin walnuts of North America which are rather soft with a tendency to warp. In the Southern Hemisphere Queensland walnut resembles true walnut in its colour and texture, although the vessels are much smaller and it possesses a very unpleasant odour when freshly cut. East Indian walnut, kokko, also has a superficial resemblance to walnut although it is coarser and harder with very large vessels so that the only real similarity is in colour. African walnut also is similar only in colour and resembles more closely a mahogany in texture, illustrating one of the problems that arises in describing woods; so often superficial resemblance is more important in a commercial description than the true botanical classification or even the properties of a wood.

American mahogany

Most of the hardwoods that have been described are derived from the temperate zones of the world, yet a very large proportion of hardwoods are obtained from tropical forests, which certainly represent the most significant potential sources of supply for the future, both for structural and decorative purposes. There was a tendency many years ago for all tropical wood to be described as mahogany but this term should correctly be confined to *Swietenia* spp., the true America mahoganies obtained from the West Indies and tropical Central America, sometimes known as Spanish mahoganies when originating from the old Spanish colonial territories such as San Domingo and Cuba. The main sources today are Honduras, Guatemala, Nicaragua and Costa Rica. American mahoganies are hard, clean and easy to work when properly seasoned, giving a good polish over an attractive golden brown colour.

African mahogany

For many years the American mahoganies have been declining in availability and, in Europe in particular, they have been largely replaced by the African mahoganies. In fact, the term mahogany has often been applied to any wood that looks superficially like American mahogany but the term African mahogany should correctly be confined to *Khaya* species which are actually related to the American *Swietenia* species. African mahoganies are often described in terms of their source, such as Lagos, Benin, Axim, Sekondi and Grand Bassam mahoganies. Many of these woods have interlocking grain which gives a particularly attractive stripe figure on radial or quarter-sawn surfaces. True *Khaya* possesses moderate to good durability but there are unfortunately a number of other West African "mahoganies" with much less consistent properties, such as the *Entandrophragma* species, which include sapele (moderately durable), utile (durable) and gedu nohor (varying from non-durable to durable). For some years gaboon, *Aucoumea klaineana,* was frequently used in the production of African "mahogany" plywood, giving rise to serious fungal decay problems in boat-building as it is non-durable; its use is now forbidden in the British Standard for marine plywood, except when preservative treated.

Even the American cedars, *Cedrela* spp., are sometimes described as mahoganies, a rather interesting situation as the African mahoganies were often in the past described as cedars, simply through the similarity in colour. One group of "mahoganies", the *Shorea* species, must be covered in greater detail because of their considerable economic importance. These woods from South East Asia are now rarely described as mahoganies but instead as meranti when derived from Malaysia, Sarawak, Brunei and Indonesia, as seraya from Sabah, and as lauan from the Philippines, although lauan also includes *Parashorea* and *Pentacme* species. These include moderately light weight, pale red or pink woods described as light red meranti, light red seraya and white lauan. Medium weight dark red woods are described as dark red meranti, dark red seraya and red lauan respectively. In addition the *Shorea* spp., Richetia group, include medium weight yellow woods described as yellow meranti or yellow seraya, whilst medium weight white or pale coloured woods, *Shorea* spp., section Anthoshorea, are described as white meranti. There are also heavy red woods of this genus available from Sarawak and Brunei (alan), from Sabah (red selangan batu), and from West Malaysia (red balau). Finally the *Shorea* species include heavy yellow or brown woods from India (sal), from Thailand (chan), from West Malaysia (balau) and from Sarawak, Brunei and Sabah (Selangan batu). It will therefore be appreciated that the Shorea genus is particularly important in South West Asia, both in respect of local use and exports throughout the world.

Meranti, Seraya, Lauan

Teak

South West Asia is also the source of teak, *Tectona grandis*. This is perhaps the most versatile and valued of woods as it combines high strength with excellent durability and a very pleasing appearance which takes polishes and oil finishes exceptionally well. Teak, a fairly heavy golden brown wood, is obtained from Burma, India and Java. It is widely used throughout the world for all purposes, varying from decorative furnishings to laboratory benches, where its resistance to fire and corrosion is particularly important, greenhouses, ship and boat-building, and heavy constructional purposes such as docks, wharfs and bridges. Burmese or Rangoon teak is strongest and also available in very large sizes which make it particularly suitable for heavy construction work. Teak from India and Java is milder, yet more decorative and thus more suitable for furnishings. Because of the high value of teak there have been many attempts to offer substitutes from other forests but none of these equal the true teak in strength and durability. The most important substitutes are iroko or Nigerian teak, *Chlorophora excelsa,* a lighter and rather coarser wood from East and West Africa, and afrormosia, *Pericopsis elata,* from West Africa.

African teak

5.2 Wood products

Decorative veneers

Veneer cutting, or the production of thin sheets of wood, originated as a means for providing furniture with an attractive wood finish at a much lower cost than that involved in using solid decorative wood. Veneers were originally prepared by slicing but most of the production is now by rotary cutting or peeling of logs as this produces veneers of very large size and thus reduces the amount of joining that is necessary. Some veneers are still prepared by slicing, principally those in which the most attractive decorative effect only occurs on a radial face, such as silver figure in oak and stripe figure in African mahogany. These decorative veneers were originally supported on solid wood, although it is obvious that the backing material must be reasonably stable and, in addition, the grain in the veneer must be at right angles to that in the backing wood. In recent years veneers have been used in the manufacture of plywood, each veneer being at right angles to the adjacent veneers so that, with the aid of modern high-strength adhesives, the cross-sectional movement can be restrained by the stable longitudinal dimension of the adjacent veneers. Whilst decorative woods are frequently used for the face veneers they are obviously unnecessary for the core veneers or indeed for any of the veneers if the plywood is to be used for a structural rather than decorative purpose.

Plywood

It is impossible to list all the woods that are used as decorative veneers or in the production of plywood. Generally structural plywoods are produced from the most convenient local source of wood but decorative veneers may be obtained from any part of the world. At one time logs were exported to industrial countries for processing but plywood mills are now situated in the forest areas, even in tropical countries. Where plywood is to be used in relatively constant interior conditions the wood species and adhesive are not particularly critical, but for external situations, such as cladding and boat-building, the situation is rather different. Specifications normally require that naturally durable species should be employed or alternatively an adequate preservative treatment should be applied. In fact, thin veneers are relatively easy to treat, even if the species is relatively impermeable as solid wood. In the past some plywoods have failed because the face veneers have been durable, yet the core veneers have been non-durable. In addition, plywood for exterior use must be manufactured using a suitable adhesive, a further point that is generally mentioned in appropriate specifications, but it is not generally appreciated or specified that exterior plywood should be manufactured from veneers with a low movement across the grain. It is true that plywood will generally restrain movement, particularly with modern high-strength adhesives, but fluctuating conditions will eventually lead to delamination, not through the failure of the adhesive but through rupturing of the wood itself.

A few structural plywoods can be mentioned. Birch, the principal wood used for plywood manufacture in the Baltic area, produces a smooth, mild but non-durable plywood which is unsuitable for exterior use unless treated in the veneer stage with a preservative. Birch is limited in availability and there has been a tendency in very recent years for at least a proportion of the Baltic plywood, particularly supplies from Finland, to be manufactured from spruce, either completely or from mixed birch and spruce veneers. In North America Douglas fir plywood is certainly most important as it possesses good natural durability which becomes quite exceptional when treated with suitable preservatives, so that it can be used for the construction of foundations of buildings. Many other woods are used for plywood manufacture in Canada and the United States but they generally lack the advantages of Douglas fir plywood and, as a result, they are rarely exported to other parts of the world. African hardwoods are used as both decorative face veneers and for complete manufacture of plywood, both within the tropical forests, and in other countries, as a result of the exporting of logs or cut veneers. There has been a tendency in the past to use non-durable species as core veneers but this is now excluded from all specifications for exterior grade plywood. In Australia many of the local Southern Hemisphere species, such as the Eucalypt spp., are widely used in plywood manufacture, but decorative woods are imported from Indonesia, Papua, New Guinea and even Africa.

Processed plywoods

In some circumstances strength is required preferentially in one direction, usually when a panel is required for a load bearing purpose, and this is most readily achieved by a central core of strips of wood to give blockboard, battenboard and laminboard; see Figure 2.17. There is little to choose between good quality plywoods, except in terms of their weather resistance, durability and appearance, the latter depending largely upon the availability of suitable face veneers. There is therefore a tendency for manufacturers to produce special plywoods in an attempt to develop an advantage over their competitors and this has resulted in the production of "processed" plywoods of all types. Film-faced plywoods with a phenolic resin coating were first produced for use as concrete form-work or shuttering but it was found that the faces were advantageous for many other uses, even for internal partition walls, benches, work-tops and storage vessels. As a result heavy film-faced plywoods were produced particularly for these more critical uses. Decorative film-faced plywoods with surfaces coated with paint or glass-reinforced plastic films are available for use as internal or external wall claddings or, alternatively, plywood can be prepared with a paper bonding on the surface which provides a particularly suitable base for painting. Metal and rubber-faced plywoods are also available for industrial uses, and mineral aggregate-faced plywoods for use as external wall cladding. Sheets of plywood manufactured as curved sections, or profiled for decorative or functional purposes, can be supplied, as well as plywoods treated with various preservatives and fire retardants.

Particle boards

The range of plywood products is enormous and, without doubt, plywoods are the best wood panel products that are available. The principal alternative is the particle board, and its manufacture is described in detail in Chapter 2. Originally particle board was devised as a means for utilising waste cutter shavings but their inconsistent size and shape tended to produce a poor quality board. Manufacturers then switched to chipping fresh logs, creating a new market for logs which, together with paper pulp manufacture, seriously affected the availability of wood. In recent years there has been a tendency for more mills to become integrated so that suitable logs are converted to sawnwood whilst unsuitable small sizes and offcuts are used for the production of particle board. In addition, there have been serious attempts to use alternative materials which are completely unsuitable for conversion to solid wood, such as flax shives or twig material from scrub. There have been serious attempts to utilize bark, both in the production of particles and adhesive. For example, in South Africa large areas were planted with wattle to produce tannin for the leather industry and there are now attempts to utilise these plantations to produce woody particles and adhesive for particle board manufacture.

Particle boards are available in various types. Whilst many boards are marketed as general purpose products, it is normal to distinguish interior structural with improved strength from interior non-structural boards. Exterior structural or non-structural boards are basically similar, except that they have been manufactured using suitable adhesives combined with durable or preserved wood chips and finally a chemical treatment that will prevent the swelling and thus disintegration of individual chips. In fact, this latter requirement is difficult to achieve and true exterior boards are not normally available, but the critical adhesive and durability properties are also required in a number of other applications where there may be a danger of fungal decay through perhaps condensation, and boards are now available which meet these requirements. Indeed, in 1970 it became mandatory in West Germany for all roof-lining particle board to be protected against decay, as well as being manufactured using a suitable adhesive; in effect this means that urea-formaldehyde boards cannot be used in certain applications as the adhesive is attacked by bacteria or fungi and has poor resistance to high moisture content, so that phenol- or resorcinol- formaldehyde adhesives must be used.

In addition to basic performance requirements particle boards are described as being single, two, three or multi-layer types, or alternatively produced by a graded density system; the purpose of these layer systems is normally to include large, coarse particles in the core of the board in order to give good strength, but fine particles in the face so that the board will give a good finish. A fine surface is essential not only for finishing with paint and varnish but also where particle boards are veneered; large chips may move in cross-section with changes in atmospheric relative humidity, causing unsightly shadowing through the veneer. Generally boards are marked with the manufacturer's name together with the specification to which they are produced so that, as with stress-graded wood, it is relatively simple to check whether an individual board is suitable for a particular purpose. Processed boards are produced, as with plywoods, but not in such an extensive range as the preferred uses for particle board are rather more limited than for plywood.

Fibre boards

All fibre boards are manufactured by a similar process to produce three basic types, varying in density through the degree of compression applied during manufacture. The lowest density board is known as insulation board and is available as sheets or tiles which are used for ceilings or soft wall surfaces; this board is often known as soft board. Medium board has a rather higher density and standard hardboard is the highest density product available. In addition, standard hardboard is frequently manufactured with oil additives and is then known as tempered hardboard; this treatment gives greater stability when the board is used as a painted exterior cladding. A variety of processed boards are available, such as perforated board and various surfaced boards which are painted, varnished or finished with various plastic or textile laminates. In addition board is available with laminates incorporating a wood grain effect or, alternatively, embossed and moulded hardboard is available, often with a wood grain effect. It is difficult to understand why wood has to be destroyed and then refabricated to give a psuedo-wood effect but this peculiarity presumably results from unrealistic commercial pressures.

5.3 Processed wood

Compreg

Impreg

A variety of products are available in which wood has been processed to improve its natural properties. Compreg consists of veneers impregnated with a phenolic resin before being compressed and polymerised at high temperature. Impreg is basically similar except that the wood is not compressed so that a greater volume of resin is used to achieve a high density. Both these products are normally employed for the production of patterns and other products in which considerable stability or hardness is required, such as in aircraft propeller blades and moulds. Staybwood is heat-

Staybwood stabilised wood, prepared by drying to about 6% and then heating to about 175 °C which causes the wood to loose its hygroscopic properties so that it becomes permanently stable. Staybwood is sometimes compressed to produce additional hardness.

Staypac In Staypac veneers are processed in a similar way but compressed to produce a very hard surface veneer. These examples of the original processed wood products were developed some years ago but more recently a relatively large number of products have become available, particularly for use as cutlery handles, and also as floor blocks as they give exceptional wearing properties. The main advantage of a resin impregnated block as a flooring is the permanent polish that is obtained, whereas normal varnish surfaces tend to break through where wear is greatest. Generally these floor blocks are prepared by impregnating with monomers of low viscosity which readily penetrate wood and then polymerising by using a chemical catalyst, heat or radiation techniques.

Plate 5.1 Traditional wood paving. The end-grain surface gives good resistance to compression loads and wear, yet a quiet and resistant surface. The end-grain blocks are also readily impregnated with creosote. Wood paving was once used extensively for urban roads but it is now used for industrial floors where its hard wear and resilience are particularly advantageous. *(Orban Bois S.A.).*

Plate 5.2 Hexagonal floor blocks, prepared by impregnating end-grain pine slices. *(Orban Bois S.A.)*

Appendix 1. Botanical Classification of Principal North Temperate Timber Trees

(a) Trees within the plant kingdom

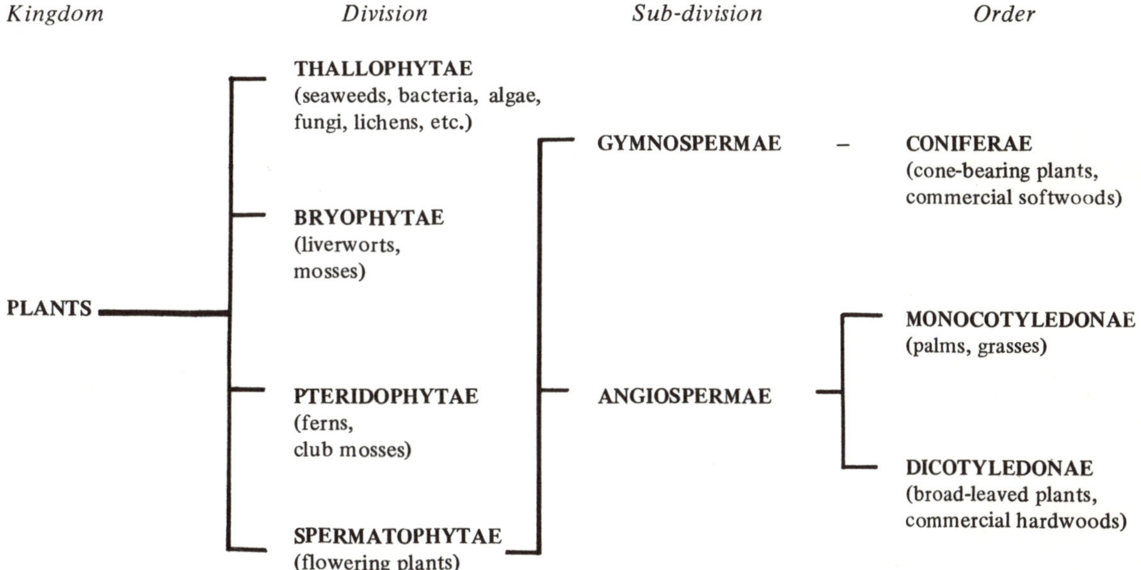

(b) Cone-bearing plants—commercial softwoods

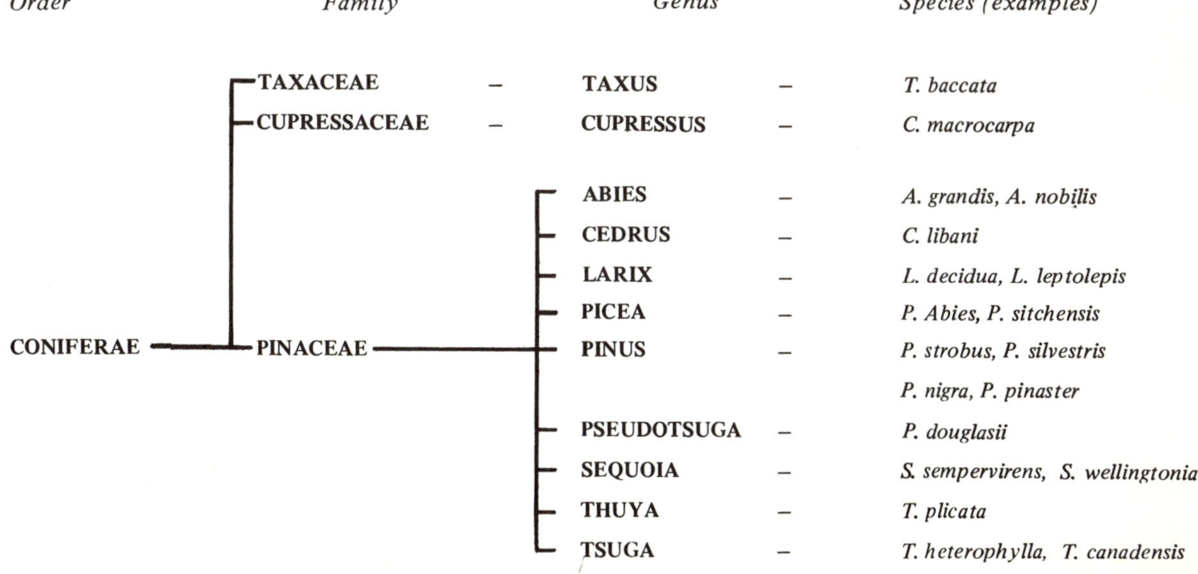

(c) Broad-leaved plants—commercial hardwood

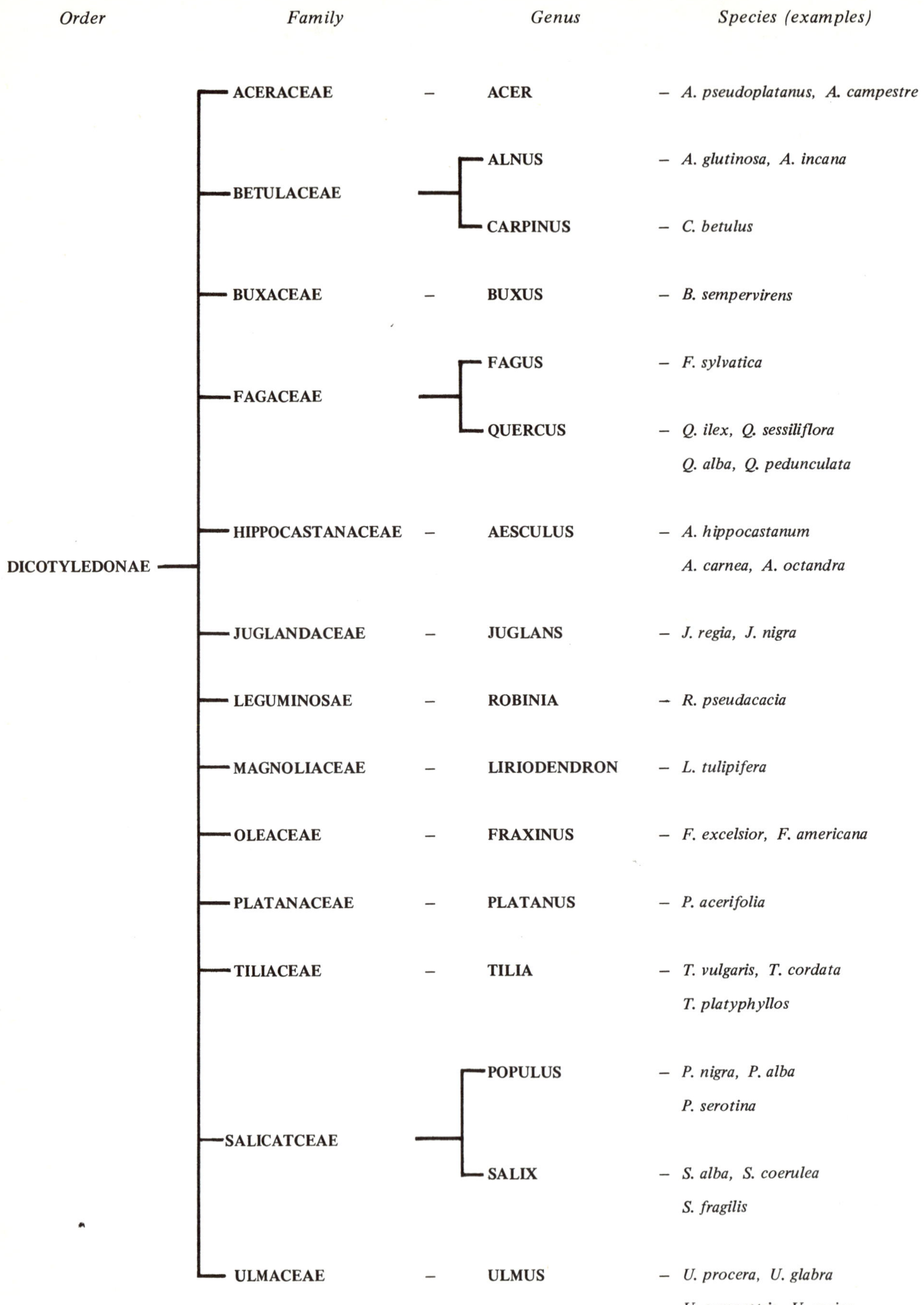

Appendix 2. Wood Structure

(a) Microscopic features

In order to examine the microscopic features of a piece of wood it is necessary either to macerate the sample or to prepare thin sections. For maceration the sample is first treated with chromic acid in order to dissolve the middle lamella and thus release the individual components. Thin sections are prepared by soaking the wood until it is soft and then cutting the sections with a microtome. In both cases stains can be used which have an affinity for various individual components so that they are more readily visible under the microscope. Maceration has the advantage that it is possible to examine individual components in their entirety but the disadvantage that their relative positions are completely unknown. Thin sections give an indication of the relative positions but it is necessary to prepare a large number of serial sections in order to encounter individual components and thus construct a three-dimensional picture of the entire wood.

This Appendix is not intended to be a comprehensive account of the microscopic features of wood and their value in the identification of individual species but simply a contribution to the understanding of the structure of wood. As an example Scots pine, *Pinus sylvestris,* has been selected as representing softwoods whilst European oak, *Quercus robur,* represents hardwoods.

The maceration of Scots pine gives a large number of long, thin, needle-like units, about 1/32″ (0.8 mm.) long. These are the principal longitudinal structural elements or tracheids, and they are hollow, four-sided and pointed at each end. "Pits" are scattered along two opposite sides of the tracheids; the bordered pits are circular whilst the simple pits are rectangular. It is also possible to distinguish parenchyma cells in the macerated sample, each oblong, box-shaped and about 1/200″ (0.1 mm.) long.

Figure A2.1 Reconstruction of a ⅛″ (3 mm.) cube of Scots pine, *Pinus sylvestris*.

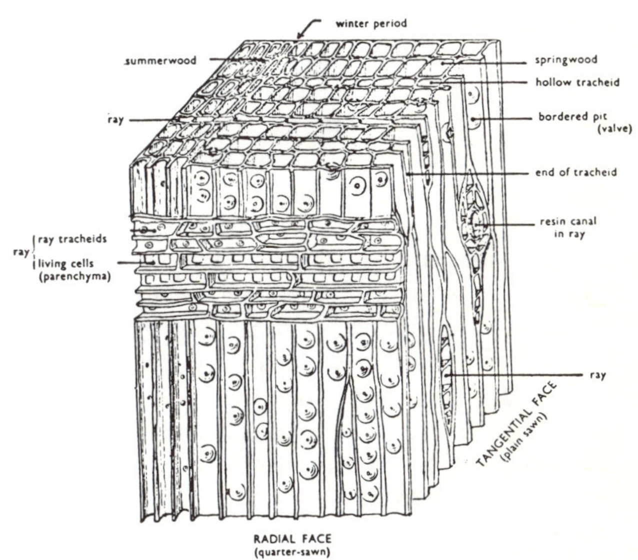

Sections must be prepared in three dimensions; cross-section, radial and tangential. Examination of the sections shows that Scots pine is predominantly composed of tracheids which were laid down in regular rows as they were formed by the cambium at the outer limit of the sapwood. The tracheids are fitted end-to-end with an overlap to give both strength and continuity for the longitudinal conduction of fluids. The spring wood tracheids are large in cross-section and thin walled compared with the summer wood tracheids which are distinguishable principally by their very thick walls. The only other longitudinal features are vertical resin canals which appear as spots in cross-section and as fine white hair lines in the radial and tangential sections. These resin canals are narrow tunnels lined with small rectangular cells. The other principal features are the medullary rays running as horizontal ribbons in the radial direction; by luck they might be shown in a radial section but the ends of the rays will certainly appear in a tangential section. In depth they consist of 3 - 10 small oblong cells with their length in the horizontal direction. The top and bottom rows consist of ray tracheids with walls of irregular thickness whilst the middle rows are parenchyma cells which are connected to the vertical tracheids through the simple pits. The rays may also incorporate horizontal resin canals.

The structure of a hardwood such as oak is entirely different. In most hardwoods the vessels are the dominant features and, in cross-section, they appear in oak as large pores. These vessels run for a distance of many feet vertically within the tree and consist of the many squat tubular cells that can be seen in the macerated preparation. These cells possess thin walls so that increasing porosity necessarily leads to decreasing strength in a hardwood. The tubular cells also possess numerous pits connecting with the adjacent tracheids and fibres. The tracheids are rather like those in softwoods but less regular in form. Fibres occur in clumps and are responsible for the principal longitudinal strength of the wood. Each fibre is spindle-shaped, long, thin and tough with a thick wall and only a small cavity. The fibres are interlocked or cemented to each other to give a hard tough wood. In addition there are small vertical rays and very large horizontal medullary rays, usually composed entirely of regular sized parenchyma cells in hardwoods and without the ray tracheids or resin canals associated with medullary rays in softwoods. In oak there are two sizes of medullary ray, one broad and the other narrow and barely visible to the naked eye. These medullary rays are a source of weakness as the ray cells are weak and the vertical tracheids and fibres are deflected around them, so that shrinkage in oak and other hardwoods is often associated with splitting through the medullary rays. In some hardwoods the medullary rays give a regular pattern, the ripple marks or silver figure frequently seen in quarter-sawn oak.

One of the features of the hardwood vessels is the formation of tyloses as the sapwood is converted into heartwood. Living cells bulge through the pits to give the appearance initially of

Figure A2.2 Reconstruction of a $\frac{1}{8}''$ (3 mm.) cube of European oak, *Quercus robur*.

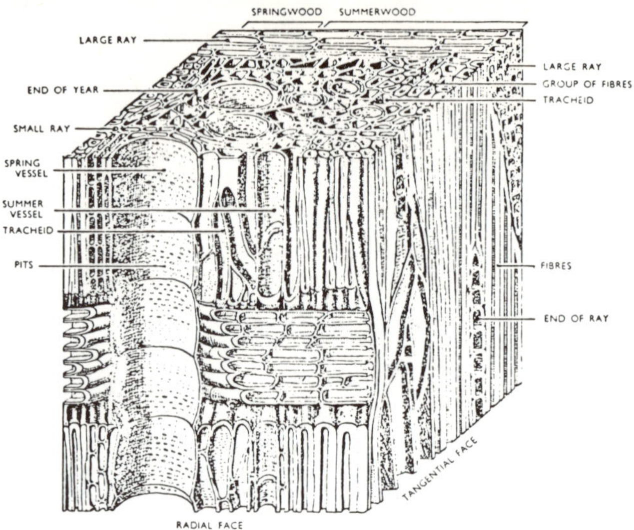

small balloons in the vessel cavities. These grow until eventually the vessel is completely blocked. Tylosis occurs only in certain species; white oak possesses tyloses which block the vessels so that it is particularly suitable for barrel making, whereas red oak has no tyloses and smoke can actually be blown through the vessels.

The annual rings represent the amount of wood formed each year but the structure in softwoods is entirely different from that in heartwoods. In softwoods a large area of thin-walled spring cells is formed, followed as the season progresses by smaller thick-walled summer or late wood cells. In hardwoods the spring wood is formed with very large vessels and as the season progresses the density of the fibres increases, accompanied by smaller vessels and a smaller number of tracheids. In fast-grown softwoods the wood is generally of low density and inferior quality whereas in fast-grown hardwoods the wood tends to have high density and superior quality. This is, of course, only a simple generalisation; very slow-grown softwoods have excellent workability but inferior strength whilst very fast-grown hardwoods tend to lack the durability associated with slower grown wood.

(b) Ultimate structure and chemistry

Wood technology is the term often used to describe the study of wood and its utilisation but wood science is a rather better description of the amalgam of botany and chemistry that is involved in a study of the ultimate structure of wood. The botanist is concerned with the gross structure of a tree, the root, stem and leaf, whilst the wood technologist concentrates on the stem alone as this is the source of the wood of commerce. However, both the botanist and the wood technologist study the histology or microscopic structure of wood, observing individual cells and identifying fibres or tracheids, parenchyma, ray cells, xylem, phloem and the other features described in Chapter 1 and in the first part of this Appendix. A study of the fine structure of wood is, however, concerned with sub-microscopic features which have only been studied in comparatively recent years since the introduction of the electron microscope. The microscopic fibres are now found to consist of orientated microfibrils and it has been deduced that these in turn consist of bundles of cellulose chains. Crystalline and amorphous cellulose structures have been identified as well as related carbohydrates such as hemicellulose and starch, all these components assembled from sugar or monsaccharide molecules. All the characteristic features of wood are found to be derived from these sugar-based structures and it is therefore hardly surprising that the entire purpose of the tree is to support and supply the leaves where these sugars are generated through photosynthesis. Thus the fine and chemical structure of wood is directly related to the gross structure of the tree; the tree is designed to support the leaves which produce the sugar which is formed into cellulose which strengthens the stem to support the leaves!

The trunk or stem of the tree is, of course, the wood of commerce. The initial twig is represented by the central pith which is surrounded by the heartwood, consisting of cells which are so far removed from the bark that they have died and become filled with various extractives. Around the heartwood is the sapwood, the living cells of the tree which conduct water upwards to the leaves. On the outside of the sapwood is the phloem in which the sugars are conveyed from the leaves to the living cells throughout the tree, providing them with energy to sustain life, and the sugar components to form cellulose, hemicellulose and starch. The outer sapwood cells immediately beneath the phloem are known as the cambium and are distinctive as they can divide to form new cells. The cambium consists essentially of two types of dividing cell, the fusiform initials and the ray initials. The fusiform initials adjacent to the sapwood give rise to xylem or wood cells which ultimately become the tracheids of conifers or softwoods, and the fibres of dicotyledons or hardwoods. The outer fusiform initials give rise to the bark. The ray initials also produce xylem but in the form of the parenchyma and ray cells. In this way the wood is formed so that the vertical tracheids and fibres give longitudinal strength and vertical transport routes, within the tracheids in conifers and within pores surrounded by fibres in dicotyledons. In contrast the ray cells are orientated along horizontal radial paths in order to provide conducting tissue between the phloem and the cells deep within the sapwood and heartwood, apparently conducting extractives towards the heartwood, sugars to the living cells in the sapwood and also often providing tissue for the storage of starch, particularly in deciduous hardwoods which shed their leaves in winter and thus require a source of stored energy to enable them to survive.

The basic physical properties of wood can be attributed to the principal components, the basic longitudinal cellular elements, the tracheids or fibres. When a fusiform initial divides, the resultant new xylem cell is surrounded by an amorphous undifferentiated substance which subsequently becomes the middle lamella. The cell rapidly achieves its ultimate length and rectangular cross-section, squeezing the middle lamella to form a thin layer between adjacent rectangular cells. The initial or primary cell wall (P layer) consists of loose, irregularly orientated microfibrils, an important feature as this thin P layer must be capable of extension as the cell grows to its ultimate dimensions. The microfibrils are orientated in a predominantly shallow spiral, perhaps at about 60° to the vertical axis of the cell. Once the P layer has achieved its ultimate dimensions the secondary wall (S layer) is commenced and is formed typically in three separate stages. The first secondary wall (S_1 layer) is thin with a predominantly shallow microfibril spiral, perhaps at about 50° to the longitudinal axis of the cell. The S_1 layer sometimes consists of two or more lamellae spiralling in opposite directions and is morphologically and structurally intermediate between the P and S_2 layers. The second secondary wall (S_2 layer) consists of the very regular closely-packed microfibrils at a very steep spiral angle, perhaps only 10 - 20° relative to the longitudinal axis, and it is also very thick, being the dominant cell wall. Finally a third secondary wall (S_3 layer) is sometimes formed, consisting of a thin, shallow orientated layer of microfibrils, perhaps at about 50° to the longitudinal axis. In all cases the secondary walls are more regularly orientated than the primary wall.

Figure A2.3 Cell wall layers in a softwood tracheid.

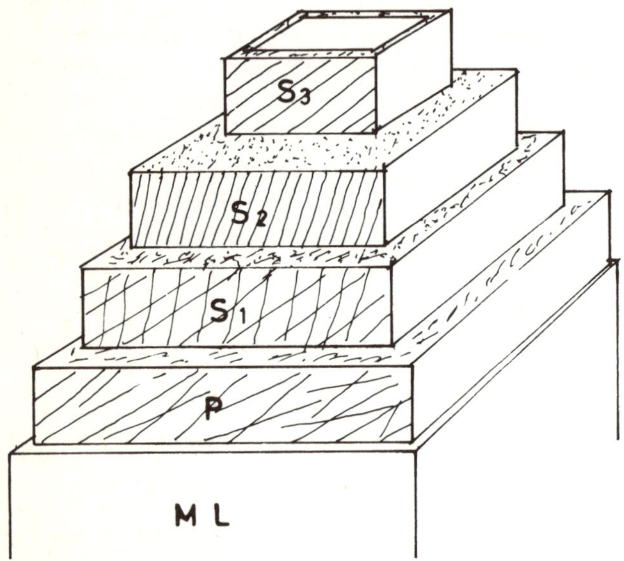

These cell walls account for the cellulose which comprises 60% of the wood substance. The S_2 layer is always dominant, and it is therefore hardly surprising to find that the basic longitudinal orientation of the microfibrils, and thus the cellulose chains, within this layer accounts for many of the basic physical properties of wood. This S_2 layer is also rather more massive in late, (summer) wood than in early (spring) wood, again accounting for the different properties between these sections of the annual rings. It was this highly orientated or crystalline structure of the cellulose chains within the cell that enabled the structure to be elucidated many years ago using crystallographic techniques such as X-ray defraction and light polarisation.

The rectangular fibres and tracheids that are so readily observed when a cross-section of a piece of wood is examined under the microscope are, of course, comparatively large but their component microfibrils are quite small. It is difficult to imagine microfibrils without a knowledge of the Angstrom (Å), the unit of dimension that must be used at this scale. By definition an Angstrom is 1 x 10^{-8} cm, or 0.0000001 mm. Once the definition of the Angstrom is appreciated it can be said that a microfibril consists of one or more elementary fibrils, each fibril having a diameter of about 35 Å, so that microfibrils have typical diameters in units of 35 Å, i.e. 35, 70, 105, 140, etc. In some cases microfibrils in wood are flattened in one direction, such as 100 x 50 Å. If these measurements are now converted to something familiar, such as a fraction of a millimetre, it will be appreciated that the elementary fibrils are very small, yet each consists of a bunch of about 40 cellulose chains, and thus the basic structure of the wood cell has been reduced to its ultimate chemical components.

It is usually considered that the cellulose chains consist of between 200 and 2,000 glucose units, although it is sometimes reported that as many as 15,000 units may be involved. As each glucose unit has a length of about 5 Å the chains are relatively long, perhaps 10,000 Å (0.001 mm) or perhaps far longer. As wood is a natural material some discontinuity of the structure occurs but the chains lie parallel for considerable lengths, perhaps over 120 units (600 Å) or more, and it is these essentially parallel cellulose chains in the dominant S_2 layer that account for the principal properties of wood.

Figure A2.4 Glucose and the formation of cellulose.

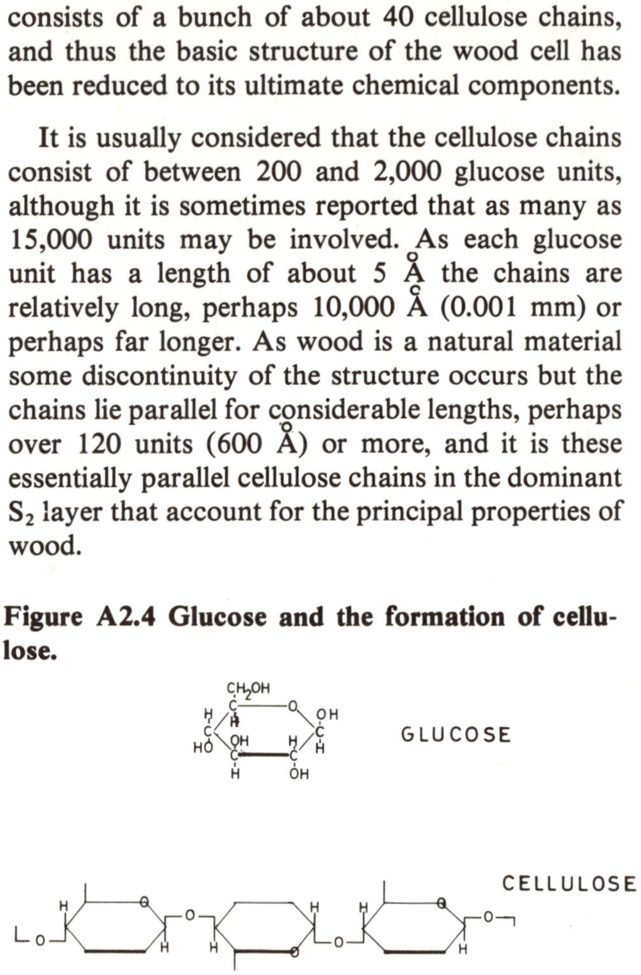

GLUCOSE

CELLULOSE

Figure A2.4 illustrates the way in which glucose units are assembled to form chains and thus the microfibrils and the principal structural elements of wood. This structure has considerable importance in determining the properties of wood. Firstly the sugar units are formed from water and carbon dioxide within the leaves by the process known as photosynthesis:

$$6H_2O + 6CO_2 \longrightarrow C_6H_{12}O_6 + 6O_2$$

water, carbon dioxide → glucose, oxygen

The water is obtained from the surrounding soil by the roots and is then conveyed up the tree through the living xylem cells (sapwood) to the leaves. Carbon dioxide is absorbed from the atmosphere by the leaves and the glucose product is conveyed down the trees in the phloem between the xylem and the bark. This production of glucose is peculiar to the process known as photosynthesis and it can only occur in the presence of the catalyst chlorophyll, the green pigment in leaves, and is dependent upon the absorption by the leaves of light, particularly ultra-violet, energy. The glucose product thus has a far higher energy level than the water and carbon dioxide constituents, and this energy can be released in a variety of ways. For example, animals may eat glucose or other sugars, converting them back to the original water and carbon dioxide by the addition of oxygen but releasing energy in the process. This energy is used for maintaining the life processes or for generating heat, and the burning of sugars is perhaps the most obvious illustration of the way in which oxygen can be combined with glucose to reverse the photosynthesis reaction and release energy. In the formation of cellulose chains a small proportion of the energy is lost but a considerable amount remains and this explains why wood burns and why it is attractive to some insects and fungi as a source of nourishment. The glucose produced by a tree is largely used for the formation of structural cellulose but the enormous mass of living cells in the roots, sapwood and leaves all require energy which they obtain from the glucose and associated sugars such as xylose which are also produced by the leaves. During darkness or the winter months the leaves are not producing glucose, yet the cells still require energy which they obtain from the glucose accumulated within the cell or, for longer periods such as winter, from deposits of starch which is formed from glucose and fairly readily reconverted to it when required. Some attacking insects are unable to utilise cellulose as a source of nourishment but they will attack wood just for the starch or simple sugar content; even mammals such as rabbits will strip the bark from trees to gain access to the sugary sap in the phloem.

It has already been explained that the cellulose chains are assembled into microfibrils which are orientated in a predominantly longitudinal direction within the cell walls of softwood tracheids and hardwood fibres. The physical properties of wood largely result from this longitudinal orientation of the cellulose chains, and this is particularly the case with the relationship between wood and water. Each glucose unit in a cellulose chain possesses three hydroxyl (–OH) groups which have an affinity for water. This ensures firstly that cellulose chains will wet easily, but in addition water will be held between the chains, pushing them apart as illustrated in the diagram. As the chains become further removed from one another the link between them becomes weaker so that a high moisture content in wood is associated with loss of strength, particularly a loss of sheer strength and rigidity so that a beam is more flexible and wood will cleave more readily when wet. If this separation of the cellulose chains were permitted to continue indefinitely the wood would eventually separate into individual chains and thus disintegrate. In fact, the movement with moisture content change can be largely attributed to the predominant orientation of the microfibrils and the cellulose chains, and thus to the S_2 layer which accounts for most of the mass of the cell wall. The swelling of the S_2 layer is limited by the extensibility of the surrounding S, and P layers.

Figure A2.5 Water absorption forcing cellulose chains apart.

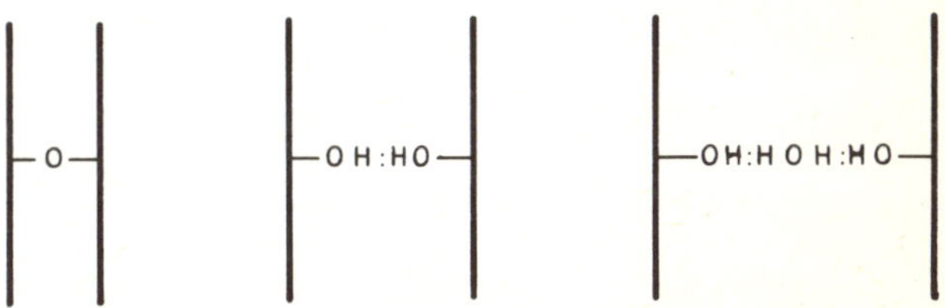

As the orientation of the microfibrils in the S_2 layer steepens from the pith to the bark of the tree the movement with change of moisture content increases approximately in proportion to the cosine of the angle of orientation. In addition it would be expected that the longitudinal movement would be proportional to the tangent of the angle of orientation, so that an angle of 10 - 20° would suggest a longitudinal movement of 17 - 36% of the cross-section movement. In fact the longitudinal movement is less than half this calculated figure, apparently due to the restraining influence of the planes of lignin in the middle lamella. The middle lamella appears to have much less control on the more massive cross-sectional movement, perhaps largely because it is orientated then as a thin envelope over the swelling material in contrast with the vertical tubes in which it obstructs vertical or longitudinal movement. However, the cross-sectional movement is still restrained when it reaches the fibre saturation point but this can be attributed largely to the very shallow microfibril angle in the P and S_1 layers which physically restrain and prevent further swelling. Prolonged water-logging introduces slow but progressive hydrolosis of the cellulose and thus weakening which eventually releases this tension and permits the S_2 layer to absorb water beyond the original fibre saturation point and swell to a greater extent; this is the case of the weakening observed in archaeological wood which has been water-logged for many centuries. Careful microscopic observation discloses that water-logging permits "ballooning" where the S_2 layer is apparently bursting through the restraining P and S_1 layers, and this damage is also observed when wood is soaked with more powerful swelling solvents such as alcohol.

Figure A2.6 Variation in moisture content of wood with changes in atmospheric relative humidity.

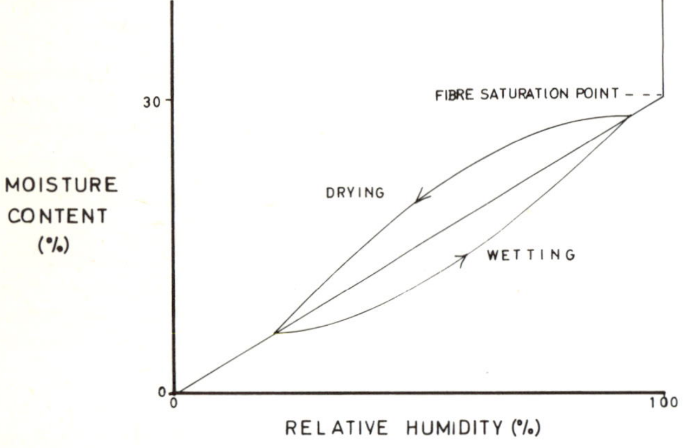

The loss or gain of water between the cellulose chains does not occur instantaneously with changes in the surrounding relative humidity but tends to lag. The reason is that changes will only occur under the influence of excess energy. In fact, energy is liberated when cellulose is wetted and this becomes apparent as the heat of wetting. This is virtually immeasurable and certainly insignificant in most circumstances, but it means that wetting will only proceed provided this heat is removed and drying will only proceed if heat is given to the wood. Whilst the dimensions of a piece of wood will depend on its moisture content this lagging effect, or hysteresis, will mean that the wood is larger than might be expected during the drying stages and smaller than might be expected during the wetting stages, as shown from the hysteresis diagram, Figure A2.6.

Only 1% of the mass of wood consists of extractives, extraneous materials and minerals such as silica, the remainder being lignin and holocellulose, the carbohydrates and related compounds derived from sugar units. The crystalline structural cellulose which is the principal component of the microfibrils is known as alpha-cellulose and accounts for about 50% of the mass of softwoods and temperate hardwoods, the remaining holocellulose, principally hemicelluloses, contributing about 25%. A further 25% is lignin, although in tropical hardwoods the lignin content is often far higher, displacing the holocellulose in proportion and giving greater rigidity but less resilience.

Appendix 3. Commercial Woods

This table lists most of the commercially important woods likely to be used for structural and decorative purposes in the temperate and neighbouring zone of the world, including woods imported from the tropics. Woods are listed alphabetically using the preferred botanical names, although cross-references are provided for alternative names. A separate entry for each principal genus gives the English, French, Dutch and German names. In addition, each species entry gives a preferred common name, followed by alternatives with the locality in which they are normally used. The sources of the species are stated, followed by notes of colour and any other important feature. In some cases species are grouped, as with *Shorea* spp., as this is the manner in which they are normally supplied commercially.

Each entry is then followed by a summary of the typical modes of use, typical uses and a brief summary of the most important properties using a numerical notation. This system may appear to be rather cumbersome but it has been chosen to enable this entire reference system to be transferred to edge-punched cards which greatly increases its usefulness.

Using the table or the card index

1. To find the entry for an individual wood check the Index to Common Names, page 199, in order to obtain the correct botanical name.
2. To find a suitable alternative wood for an individual use prepare a code for the required properties and then search the table for a match. To simplify this type of search the table is divided into the major botanical divisions, i.e. hardwoods and softwoods, but it must be appreciated that, for example, some light hardwoods might be suitable as replacements for traditional softwoods in structural applications.
3. The codings are designed for use with a Cope-Chat Paramount sorting card P.3. These cards have edge-punchings which can be cut to characterize each property. The alphabetical punchings are used for the Genus; i.e. Baltic redwood, *Pinus sylvestris,* would have an entry at P. The modes of use, A E I O U, typical uses 1-13, properties 14-37, and botanical classification 38-39 are also entered at the appropriate points. A needle can then be placed through an appropriate punching to select the required cards, the cut punchings falling clear so that they can then be processed for the next property. This system can be infinitely extended to incorporate new woods as they become available, and it is the only realistic simple system for matching wood properties as a search of such an extensive table is very tedious.

Table and card notation

Uses

Mode of use

Solid wood	A
Laminated wood (beams)	E
Plywood	I
Particle board	O
Fibre board	U

Typical end-uses

Carcassing (framing) in buildings	1
Cladding in buildings	2
External joinery (millwork)	3
Internal joinery (millwork)	4

Floors 5
Furniture and panelling 6
Boatbuilding 7
Marine (heavy), bridges 8
Poles 9
Posts (fence) 10
Boxes, crates, pallets 11
Trucks, rail coaches, etc. 12
Railway sleepers (ties) 13

Properties

Texture

Coarse 14
Medium 15
Fine 16
Interlocked grain 17
Straight grain 18

Density (at 12-15% moisture content)

Dense (over 0.64 S.G.) 19
Medium 20
Light (under 0.48 S.G.) 21

Note: Very light (less than 0.35) and very heavy (more than 0.90) woods are noted separately, e.g. Balsa, *Ochroma*, and Lignum vitae, *Guaiacum*.

Strength (a combined assessment of strength in bending and compression along the grain)

Very high 22
High 23
Moderately high 24
Medium 25
Low 26

Movement (the sum of the percentage changes in dimension in the tangential and radial directions for a change in relative humidity from 90% to 60%)

Large (over 4.5%) 27
Medium 28
Small (under 3.0%) 29

Durability (heartwood; sapwood is invariably less durable)

Very durable 30
Durable 31
Moderately durable 32
Non-durable or perishable 33

Permeability to wood preservatives (heartwood; sapwood is often more permeable)

Very resistant 34
Resistant 35
Moderately resistant 36
Permeable 37

Botanical classification

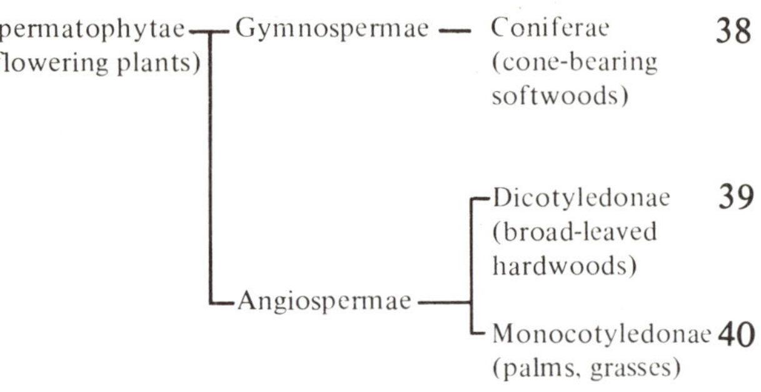

Note: 40 does not appear on the cards but is assumed when both 38 and 39 are unpunched. In fact there are no Monocotyledons in the table but this entry is necessary so that cards can be prepared for countries in which they are used.

Preparing cards

Cope-chat Paramount sorting cards ref. P.3 may be obtained in boxes of 1,000 from The Copeland-Chatterson Co. Ltd., Dudbridge, Stroud, GL5 3EU, England, together with hand-punches for cutting the card entries and sorting needles. A copy of each entry in the table is stuck to the face of the card, or each card can be retyped with the information given in the table. A Punching is made to identify the genus, i.e. P for *Pinus sylvestris*, then 38 for softwoods or 39 for hardwoods, followed by punchings for the uses and properties as shown in the table.

Examples
(a) Table entries

Khaya ivorensis

Local names: African mahogany (includes *K. anthotheca, K. nyasica*); *Local names:* Ghana, Ivory Coast, Grand Bassam, etc. mahogany, according to origin (includes *K. anthotheca*), Lagoswood, (Nigeria), ogwango (Nigeria), ngollon (Cameroons); *Source of origin:* West Africa; *Colour:* Lt., dk. red-brown

Uses		*Properties*					
Mode of use	*Typical uses*	*Texture*	*Density*	*Strength*	*Movement*	*Durability*	*Permeability*
A, I	2, 3, 4, 5, 6, 7,	15, 17, 18	20	25	29	32	34

Mansonia altissima

Common name: Mansonia: *Local names:* bètè (France, Ivory Coast), aprono (Ghana), ofun (Nigeria); *Source of origin:* West Africa: *Colour:* dk. brown-grey (purple when fresh)

Triplochiton scleroxylon

Common names: Obeche (Nigeria), wawa (Ghana); *Local names:* Arere (Nigeria), ayous (Cameroons, Zaire), samba (Ivory Coast); *Source of origin:* West Africa; *Colour:* Cream-yellowish

	Uses		*Properties*					
	Mode of use	*Typical uses*	*Texture*	*Density*	*Strength*	*Movement*	*Durability*	*Permeability*
Mansonia altissima	A	3, 4, 5, 6	16, 17, 18	20	24	28	30	34
Triplochiton scleroxylon	A, I	4, 6, 7, 12	15, 17, 18	21	26	29	33	35

b) Card entries

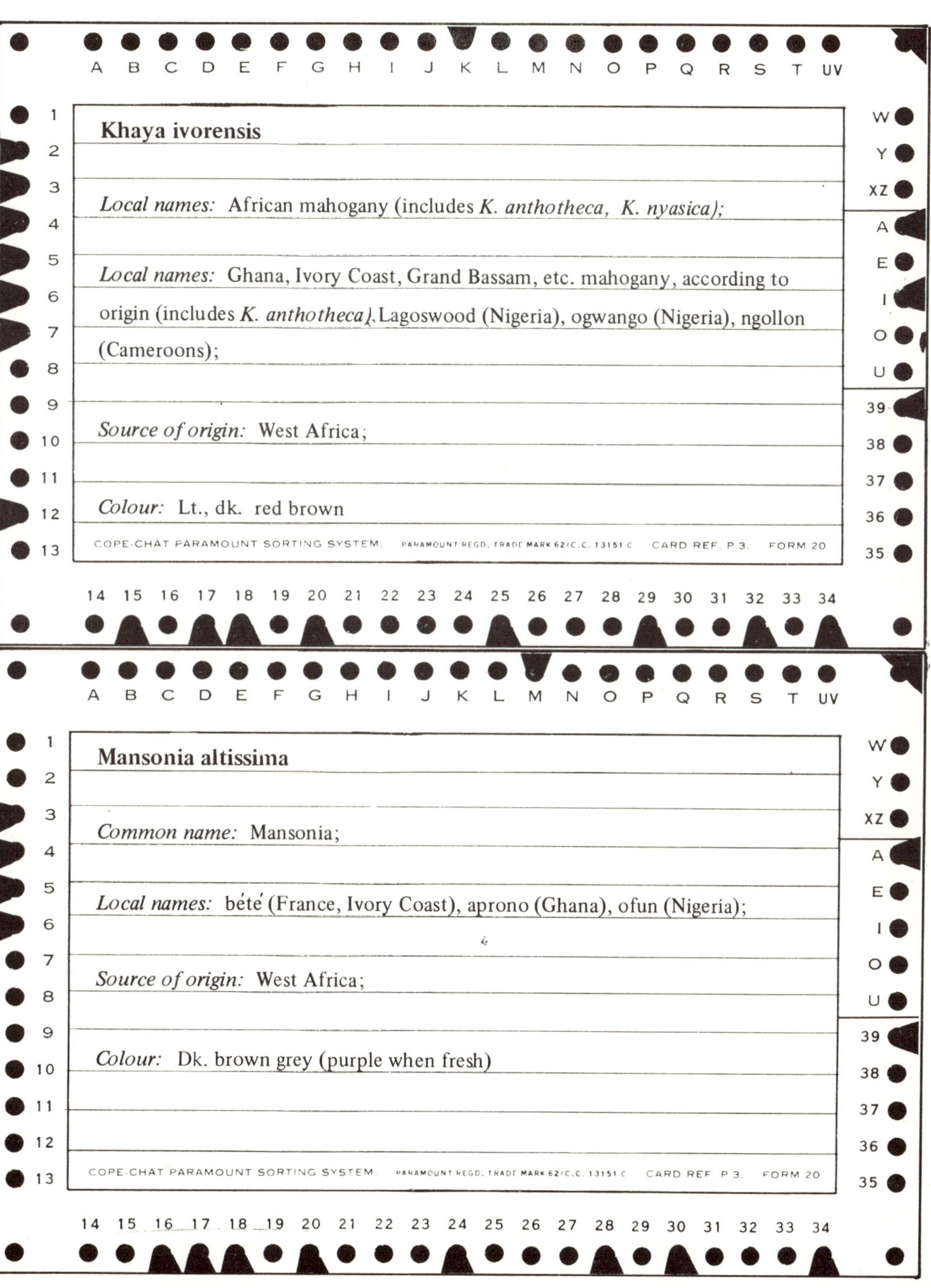

A B C D E F G H I J K L M N O P Q R S T UV

W Y XZ A E I O U 39 38 37 36 35

Khaya ivorensis

Local names: African mahogany (includes *K. anthotheca, K. nyasica);*

Local names: Ghana, Ivory Coast, Grand Bassam, etc. mahogany, according to origin (includes *K. anthotheca),* Lagoswood (Nigeria), ogwango (Nigeria), ngollon (Cameroons);

Source of origin: West Africa;

Colour: Lt., dk. red brown

COPE-CHAT PARAMOUNT SORTING SYSTEM. PARAMOUNT REGD. TRADE MARK 62/C.C. 13151 C CARD REF. P.3. FORM 20

1 2 3 4 5 6 7 8 9 10 11 12 13

14 15 16 17 18 19 20 21 22 23 24 25 26 27 28 29 30 31 32 33 34

A B C D E F G H I J K L M N O P Q R S T UV

W Y XZ A E I O U 39 38 37 36 35

Mansonia altissima

Common name: Mansonia;

Local names: bété (France, Ivory Coast), aprono (Ghana), ofun (Nigeria);

Source of origin: West Africa;

Colour: Dk. brown grey (purple when fresh)

COPE-CHAT PARAMOUNT SORTING SYSTEM. PARAMOUNT REGD. TRADE MARK 62/C.C. 13151 C CARD REF. P.3. FORM 20

1 2 3 4 5 6 7 8 9 10 11 12 13

14 15 16 17 18 19 20 21 22 23 24 25 26 27 28 29 30 31 32 33 34

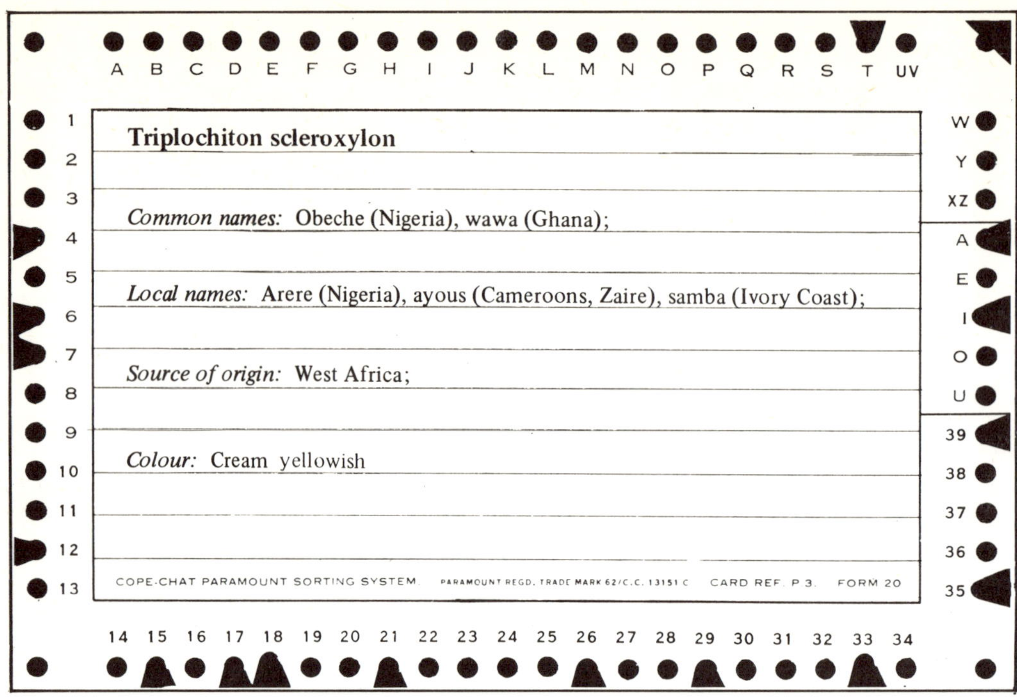

A B C D E F G H I J K L M N O P Q R S T UV

Triplochiton scleroxylon

Common names: Obeche (Nigeria), wawa (Ghana);

Local names: Arere (Nigeria), ayous (Cameroons, Zaire), samba (Ivory Coast);

Source of origin: West Africa;

Colour: Cream yellowish

COPE-CHAT PARAMOUNT SORTING SYSTEM. PARAMOUNT REGD. TRADE MARK 62/C.C. 13151 C CARD REF. P 3. FORM 20

It can be seen that, although African mahogany and mansonia are similar in many respects, their differences account for their typical uses. Mansonia is fine-grained (16) and thus is more suitable than medium-grained mahogany (15) for sill and drip mouldings when used for external joinery (millwork) (3). The colour and texture of mahogany is attractive so that it used for veneered doors and plywood face veneers (1). Mahogany with small movement (29) is used for pattern making whereas mansonia has medium movement (28) and is unsuitable. As a further contrast obeche (wawa), with medium grain (15), low density (21), low strength (26), small movement (29) and low durability (33) is used for interior joinery (4) but is unsuitable for exterior use. It is also unsuitable for load carrying, such as stairs, or where there is serious wear, such as floors.

Table of woods

	Uses		*Properties*					
	Mode of use	*Typical uses*	*Texture*	*Density*	*Strength*	*Movement*	*Durability*	*Permeability*
Softwoods—Coniferae								
Abies spp.— *Common name:* Fir (English), Sapin (French), Den (Dutch), Weisstanne (German)								
Abies alba *Common names:* Silver fir, whitewood (but also applied to *Picea abies*—spruce); *Sources of origin:* Central and South Europe, British Isles				21	24		33	36
Abies amabilis *Common name:* Amabilis fir; *Local names:* Pacific silver fir (USA), silver fir (USA), white fir (USA); *Sources of origin:* West Canada, West USA				21	26		33	35
Abies balsamea *Common name:* Balsam fir; *Local name:* Balsam (USA); *Sources of origin:* Central and East Canada, North USA; *Colour:* Cream or pale brown	A	1, 11	15	21	26		33	35

Species	Uses: Mode of use	Uses: Typical uses	Properties: Texture	Properties: Density	Properties: Strength	Properties: Movement	Properties: Durability	Properties: Permeability
Abies grandis *Common name:* Grand fir; *Local names:* Lowland fir (Canada), white fir (USA), western balsam fir (USA); *Sources of origin:* British Columbia, West USA, British Isles; *Colour:* Yellowish brown		1, 4, 11	14	21		29	33	35
Abies lasiocarpa *Common name:* Alpine fir; *Sources of origin:* British Columbia, Alberta, West USA				21				
Abies procera (A. nobilis) *Common name:* Noble fir; *Sources of origin:* West USA, British Isles; *Colour:* Yellowish brown		1, 4, 11	14	21			33	36
Agathis australis *Common name:* New Zealand kauri; *Local name:* Kauri pine (UK); *Source of origin:* New Zealand				20			32	35
Agathis dammara (A. alba) *Common name:* East Indian kauri; *Local names:* Sabah kauri (UK), Sarawak kauri (UK),bindang (Sarawak), bendang (Sarawak), mengilan (Sabah), Malaysian kauri (UK), damar minyak (Malaysia); *Sources of origin:* East Indies, Malaysia, New Guinea; *Colour:* White-pale yellow-pale brown	A	1, 5, 6	16, 18	20			33	
Agathis robusta (A. palmerstonii, A. microstachya) *Common name:* Queensland kauri; *Local names:* Queensland kauri pine (UK), kauri pine (Australia); *Source of origin:* Australia; *Colour:* Pale brown	A, I	5, 6		20	26		33	
Agathis vitiensis *Common name:* Fijian kauri; *Local names:* Fijian kauri pine (UK), dakua (Fiji); *Source of origin:* Fiji				20			33	35
Araucaria angustifolia *Common name:* Parana pine: *Local name:* Brazilian pine (USA); *Sources of origin:* Argentina, South Brazil, Paraguay; *Colour:* Pale cream	A, I	1, 4, 5, 6	16	20	24	28	33	36
Araucaria araucana *Common name:* Chile pine: *Local name:* Chilean pine (UK); *Source of origin:* Chile				20		28		
Araucaria bidwillii *Common name:* Bunya pine; *Local names:* Caribbean pine (Australia), Queensland pine (includes *A. cunninghamii*); *Source of origin:* Australia; *Colour:* Pink-brown	A, I	1, 4, 6, 11	15	20	25		32	35
Araucaria clinkii *Common name:* Clinkii pine; *Source of origin:* New Guinea; *Colour:* Pale yellow-brown	A, I	4, 5		21	25		33	
Araucaria cunninghamii *Common name:* Hoop pine; *Local names:* Queensland pine (includes *A. bidwillii*); *Sources of origin:* Australia, New Guinea; *Colour;* Cream to lt. brown	A, O	7		20	25		33	36
Athrotaxis zelaginoides *Common name:* King William pine; *Source of origin:* Australia; *Colour:* Yellow-pink	A	2, 7	16, 18	21	25		32	35
Callitris calcarata *Common name:* Black cypress pine (Australia); *Source of origin:* Australia; *Colour:* Dk. brown								
Callitris columellaris *Common name:* White cypress pine (Australia); *Source of origin:* Australia; *Colour:* Yellow to dk. brown	A	1, 2, 5,		19	25		30	

	Uses		Properties					
	Mode of use	*Typical uses*	*Texture*	*Density*	*Strength*	*Movement*	*Durability*	*Permeability*
Calocedrus decurrens (Libocedrus decurrens) *Common name:* Incense cedar; *Local name:* California incense cedar (USA); *Sources of origin:* Oregon, California; *Colour:* Reddish brown		4, 6, 10	15	21	25		32	35
Cedrus **spp.—** *Common names:* Cedar (English), Cèdre (French), Ceder (Dutch), Zedernholz (German)								
Cedrus atlantica *Common name:* Cedar (see also *C. deodara* and *C. libani*); *Local names:* Atlas cedar (UK), Atlantic cedar (UK); *Sources of origin:* British Isles, India				20			31	35
Cedrus deodara *Common name:* Cedar (see also *C. atlantica* and *C. libani*); *Local name:* Deodar (India and UK); *Sources of origin:* British Isles, India; *Colour:* Lt. brown		4	16	20		28	31	35
Cedrus libani (C. libanotica) *Common Name:* Cedar (see also *C. atlantica* and *C. deodara); Local name:* Cedar of Lebanon; *Sources of origin:* British Isles, India, E. Mediterranean				20			31	35
Chamaecyparis lawsoniana (Cupressus lawsoniana) *Common name:* Port Orford cedar; *Local name:* Lawson's cypress (UK); *Sources of origin:* Oregon, California, British Isles; *Colour:* Lt. yellow-brown		1, 4, 7, 13	14	21	25		31	36
Chamaecyparis nootkatensis (Cupressus nootkatensis) *Common name:* Yellow cedar; *Local names:* Alaska yellow cedar (USA), nootka false cypress (USA), yellow cypress (Canada), Pacific coast yellow cedar (Canada); *Sources of origin:* Pacific coast of USA and Canada; *Colour:* Yellow		4, 6, 7, 9	15	20	24		31	35
Chamaecyparis thyoides (Cupressus thyoides) *Common name:* Southern white cedar; *Local names:* Atlantic white cedar (USA), white cedar (USA); *Source of origin:* Eastern USA; *Colour:* Pink or light brown		9, 10, 11	16, 18	21				
Cryptomeria japonica *Common name:* Sugi; *Local name:* Japanese cedar (UK); *Source of origin:* Japan				21			31	
Cupressus **spp., see also** ***Chamaecyparis*** **and** ***Taxodium*** *Common name:* Cypress; *Local names:* Other names according to geographical origin, e.g. East African cypress; *Sources of origin:* Mediterranean, East Africa				21				
Dacrydium colensoi *Common name:* New Zealand silver pine; *Local name:* Westland pine; *Source of origin:* New Zealand							30	34
Dacrydium cupressinum *Common name:* Rimu; *Local name:* Red pine (NZ); *Source of origin:* New Zealand; *Colour:* Brown	A, I	1, 5, 6		20			32	35
Dacrydium elatum *Common name:* Sempilor; *Local name:* Malor; *Sources of origin:* Sarawak, Borneo				20			33	
Dacrydium franklinii *Common name:* Huon pine; *Source of origin:* Australia				20	25		31	
Fitzroya cupressoides *Common name:* Alerce; *Source of origin:* Chile				20			31	37
Juniperus procera *Common name:* East African pencil cedar; *Local name:* African pencil cedar (UK); *Source of origin:* East Africa				20			31	34

	Uses		Properties					
	Mode of use	*Typical uses*	*Texture*	*Density*	*Strength*	*Movement*	*Durability*	*Permeability*
Juniperus virginiana (J. silicicola) *Common name:* Virginian pencil cedar; *Local name:* Eastern red cedar (USA); *Source of origin:* Eastern USA; *Colour:* Purplish or red	A	4, 10	16	20			31	
Larix spp— Larch (English), Mélèze (French), Lariks (Dutch), Lärche (German)								
Larix decidua (L. europaea) *Common name:* European larch; *Source of origin:* Europe, including British Isles				20	24	29	32	35
Larix eurolepis *Common name:* Dunkeld larch; *Local name:* Hybrid larch (UK); *Source of origin:* British Isles; *Colour:* Hybrid (*L. decidua and L. kaempferi)*				21				
Larix kaempferi (L. leptolepis) *Common name:* Japanese larch; *Sources of origin:* Japan, British Isles				20	24	29	32	35
Larix laricina *Common name:* Tamarack larch; *Local names:* Tamarack (Canada, USA), Eastern larch (USA), eastern Canadian larch (UK); *Sources of origin:* Canada, North-east USA; *Colour:* Reddish brown		9, 10, 11, 13	15	20	24		32	35
Larix occidentalis *Common name:* Western larch; *Sources of origin:* British Columbia, western USA; *Colour:* Reddish brown	A	1, 5, 9, 11, 12, 13	14, 18	20	24		32	35
Larix russica (L. sibirica) *Common name:* Siberian larch; *Sources of origin:* West Siberia, North-east Russia				20				
Libocedrus decurrens, see ***Calocedrus decurrens***								
Phyllocladus rhomboidalis *Common name:* Celery-top-pine; *Sources of origin:* Tasmania, New Guinea; *Colour:* Pale cream	A	3, 4, 5, 7, 12		19	25		31	
Picea spp.— *Common names;* Spruce (English), Épicea (French), Spar (Dutch), Fichte (German)								
Picea abies (P. excelsa) *Common names:* Norway spruce, whitewood (but also applied to *Abies alba*—fir); *Local names:* Baltic, Finnish, Swedish, Russian, Yugoslavian or northern whitewood (UK), common spruce (UK), European spruce (UK), white deal (UK), white fir (UK), white pine (Scotland); *Sources of origin:* North and central Europe, British Isles; *Colour:* White	A	1, 4, 5	16	21	25	29	33	35
Picea engelmannii *Common name:* Engelmann spruce; *Local names:* Mountain spruce (Canada), Rocky Mountain spruce (Canada); *Sources of origin:* Alberta, British Columbia, west USA; *Colour:* Pale yellowish brown		1, 4, 6, 11, 12	15	21	26		33	35
Picea glauca *Common names:* Western white spruce, eastern Canadian spruce; *Local names:* White spruce (Canada and USA), Quebec spruce (UK), New Brunswick spruce (UK), Nova Scotia spruce (UK), St. John spruce (UK), maritime spruce (UK); *Sources of origin:* Canada, USA; *Colour:* Pale yellowish brown		1, 4, 6, 11, 12	15	21			33	35
Picea mariana *Common name:* Black spruce; *Source of origin:* East Canada; *Note:* Often shipped with eastern Canadian spruce *(P. glauca); Colour:* Pale yellowish brown		1, 4, 6, 11, 12	15	20			33	35

Picea rubens (P. rubra)

Common name: Red spruce; *Source of origin:* East Canada; *Note:* Often mixed with eastern Canadian spruce (*P. glauca*); *Colour:* Pale yellowish brown

Picea sitchensis

Common name: Sitka spruce; *Local names:* Silver spruce (UK, Canada, USA), tideland spruce (Canada, USA); *Sources of origin:* Western Canada, west USA, British Isles; *Colour:* Pale yellow-brown

Pinus spp.—

Common names: Pine (English), Pin (French), Pijn (Dutch), Kiefer (German)

Pinus banksiana

Common name: Jack pine; *Local names:* Princess pine (Canada), Banksian pine (Canada); *Sources of origin:* East and central Canada, north USA; *Colour:* Lt. brown

Pinus brutia

Common name: Aleppo pine

Pinus caribaea

Common name: Caribbean pitch pine (also applied to *P. oocarpa); Local names:* Caribbean longleaf pitch pine (UK), Nicaraguan pitch pine (UK), etc. depending upon source, Bahamas pitch pine, bunya pine (Australia); *Sources of origin:* Central America, Cuba, the Bahamas, Australia; *Colour:* Pink-brown

Pinus contorta

Common name: Lodgepole pine; *Local name:* Contorta pine (UK); *Sources of origin:* British Columbia, Alberta, West USA, British Isles; *Colour:* Pale yellowish brown

Pinus echinata

Common name: Shortleaf pine; *Local names:* American pitch pine (UK), Gulf coast pitch pine (UK), southern (or southern yellow) pine (USA) (these alternatives include other species); *Source of origin:* South USA: *Colour:* Yellowish brown

Pinus elliottii

Common name: Slash pine; *Local names:* American pitch pine (UK), Gulf coast pitch pine (UK), southern (or southern yellow) pine (USA) (these alternatives include other species); *Sources of origin:* USA, Australia; *Colour:* Yellow-red-brown

Pinus lambertiana

Common name: Sugar pine; *Local names:* Californian sugar pine (USA), soft pine (UK); *Sources of origin:* Oregon, California; *Colour:* Lt. brown

Pinus monticola

Common name: Western white pine; *Local names:* Idaho white pine (USA), soft pine (UK); *Sources of origin:* West Canada, west USA: *Colour;* Cream to lt. brown

Pinus nigra, var. *nigra (P. laricio,* var. *austriaca)*

Common name: Austrian pine; *Local name:* Bosnian pitch pine; *Source of origin:* South-east Europe (British Isles)

Pinus nigra, subsp. *laricio (P. nigra,* var. *calabrica)*

Common name: Corsican pine; *Sources of origin:* South Europe, British Isles

Pinus oocarpa

Common name: Caribbean pitch pine (also applied to *P. caribaea*); *Local names:* Caribbean longleaf pitch pine (UK), Nicaraguan pitch pine (UK), etc. according to source, west coast Nicaraguan pitch pine (UK), ocote pine (Central America); *Sources of origin:* Central America, Cuba, the Bahamas

	Uses		*Properties*					
	Mode of use	*Typical uses*	*Texture*	*Density*	*Strength*	*Movement*	*Durability*	*Permeability*
Picea rubens		1, 4, 6, 11, 12	15	21			33	35
Picea sitchensis	A	1, 2, 4, 5, 11, 12	15	21	25	29	33	35
Pinus spp.								
Pinus banksiana		9, 10, 11, 13		20	25		33	36
Pinus brutia						29		
Pinus caribaea	A	1, 2, 3, 4, 5	15	19	24	28	33	36
Pinus contorta		9, 10, 13	15	21	26	29	33	35
Pinus echinata	A, U	1, 9, 11, 13	15	19	24		33	37
Pinus elliottii	A, I, U	1, 4, 5, 9, 11, 13	15	19	25		33	
Pinus lambertiana		1, 2, 3, 4, 11	14	21	26	29		
Pinus monticola	A	1, 2, 3, 4, 11, 12	15	21	25		33	36
Pinus nigra, var. *nigra*				20		29	33	
Pinus nigra, subsp. *laricio*				20	24	29	33	37
Pinus oocarpa				19	23	28	33	36

Species	Uses: Mode of use	Uses: Typical uses	Properties: Texture	Properties: Density	Properties: Strength	Properties: Movement	Properties: Durability	Properties: Permeability
Pinus palustris *Common names:* Longleaf pine or longleaf yellow pine (UK), American pitch pine (UK), Gulf coast pitch pine (UK), southern (or southern yellow)pine (USA) (these alternatives include other species); *Source of origin:* South USA; *Colour:* Yellow-brown	U	1, 2, 3, 4, 5, 6, 9, 11, 12, 13	15	19	23	28	33	36
Pinus patula *Common name:* South African pine; *Source of origin:* South Africa	A	1, 4, 5, 11		21	26		33	37
Pinus pinaster *Common name:* Maritime pine; *Sources of origin:* South and south-west Europe				20	24		32	35
Pinus ponderosa *Common name:* Ponderosa pine (may include *P. jeffreyi*); *Local names:* Western yellow pine (USA, Australia), British Columbia soft pine (Canada), Californian white pine (USA); *Sources of origin:* British Columbia, west USA; *Colour:* Reddish brown	A	1, 4, 6, 9, 10, 11	14	21	24		33	36
Pinus radiata (P. insignis) *Common name:* Radiata pine; *Local names:* Monterey pine (USA, Australia), insignis pine (S. Africa); *Sources of origin:* Australia, New Zealand, South Africa; *Colour:* white-cream-yellow	A, I	2, 4, 5		20	25	28	33	36
Pinus resinosa *Common name:* Canadian red pine; *Local names:* Red pine (Canada, USA), Norway pine (USA); *Sources of origin:* East Canada, North USA; *Colour:* Lt. reddish brown		9, 10, 11, 13		20	24		33	36
Pinus sibirica (P. cembra,* var. *sibirica), (P. koraiensis) *Common name:* Siberian yellow pine; *Local names:* Siberian pine (UK), Korean pine (UK), Manchurian pine (UK), Siberian cedar (UK); *Sources of origin:* Siberia, Manchuria				21				
Pinus strobus *Common names:* Yellow pine, white pine; *Local names:* Eastern white pine (Canada, USA), northern (white) pine (USA), Quebec (yellow) pine (UK), soft pine (UK), Weymouth pine (UK); *Sources of origin:* East Canada, east USA; *Colour:* Lt. yellow-brown		1, 4, 6, 11	16	21	24	29	33	36
Pinus sylvestris *Common names:* Scots pine, redwood (Europe only); *Local names:* Red or yellow deal (UK), Baltic/Finnish/Swedish/Archangel/Siberian/Polish redwood or yellow deal, according to origin (UK), Scots fir (UK), red pine (Scotland), Norway fir (UK); *Sources of origin:* North Europe, British Isles, west Siberia; *Colour:* Pale red-brown	A	1, 3, 4, 5, 9, 12, 13	15	20	24	28	33	36
Pinus taeda *Common name:* Loblolly pine; *Local names:* American pitch pine (UK), Gulf coast pitch pine (UK), southern (or southern yellow) pine (USA) (these alternatives include other species), Queensland field pine (Australia); *Sources of origin:* South USA, Australia; *Colour:* Yellow-red brown	A, U	1, 5, 9, 11, 13	15	20	25		32	35
Podocarpus amarus *Common name:* Black pine (Australia); *Source of origin:* Australia; *Colour:* Brown	A, I	4, 6		21	26		33	
Podocarpus dacrydioides *Common name:* New Zealand white pine; *Local names:* White pine (NZ), kahikatea (NZ); *Source of origin:* New Zealand				21			32	35
Podocarpus elatus *Common name:* Brown pine; *Source of origin:* Australia; *Colour:* Brown	A	1, 3, 4, 6		20			32	

Species	Uses		Properties					
	Mode of use	Typical uses	Texture	Density	Strength	Movement	Durability	Permeability
Podocarpus ferrugineus *Common name:* Miro; *Source of origin:* New Zealand				20			32	35
Podocarpus spp., including *P. gracilior, P. milanjianus, P. usambarensis* *Common name:* Podo; *Local name:* Yellowwood (S. Africa); *Sources of origin:* Kenya, Tanzania, Uganda; *Colour:* Yellowish	A, I	4, 5	16	20		28		
Podocarpus guatemalensis *Common name:* British Honduras yellowwood; *Local name:* Cypress (British Honduras); *Source of origin:* British Honduras				20		29	31	36
Podocarpus nubigenus *Common name:* Manio (includes other species); *Local name:* (manilihuan) maniu (Chile); *Source of origin:* Chile				20			31	
Podocarpus salignus (P. chilinus) *Common name:* Manio (includes other species); *Local name:* Maniu (de la frontera) (Chile); *Source of origin:* Chile				20			31	
Podocarpus spicatus *Common name:* Matai; *Local names:* Mai (NZ), black pine (NZ); *Source of origin:* New Zealand; *Colour:* Pale cream	A, I	2, 3, 5,		20			32	35
Podocarpus totara (P. hallii) *Common name:* Totara; *Source of origin:* New Zealand				20				
Pseudotsuga menziesii (P. taxifolia, P. douglasii) *Common name:* Douglas fir; *Local names:* (British) Columbian pine (UK), Oregon pine (USA); *Sources of origin:* British Columbia, west USA, British Isles; *Colour:* Pink-brown	A, E, I	1, 2, 3, 4, 5, 8, 9, 11, 13	15	20	24	29	33	35
Saxegothaea conspicua *Common name:* Manio (with *Podocarpus nubigenus* and *salignus); Local name:* Maniu (macho) (Chile); *Source of origin:* Chile								
Sequoia sempervirens *Common names:* Sequoia, redwood (USA only); *Local name:* Californian redwood (UK, USA); *Source of origin:* USA; *Colour:* Red-brown	A,	1, 2, 3, 4, 7, 8, 10	14	21	24		31	36
Sequoiadendron giganteum (S. gigantea, S. wellingtonia) *Common name:* Wellingtonia; *Sources of origin:* California, British Isles				21				
Taxodium distichum *Common name:* Southern cypress; *Local names:* Bald, Louisiana or swamp cypress (USA) (wood from coastal swamps known as tidewater red or black cypress through dark heartwood, whereas inland wood is white or yellow cypress); *Source of origin;* South-east USA; *Colour:* Yellow to lt. or dk. brown	A	2, 3, 4, 6, 8, 9, 10, 11, 13		20			31	36
Taxus baccata *Common name:* Yew; *Local names:* Common or European Yew (UK); *Sources of origin:* Europe, British Isles				19			31	35
Tetraclinis articulata *Common name:* Thuya; *Sources of origin:* North Africa, Malta; *Note:* Used for decorative burrs; not to be confused with thuya—cedar								
Thuja occidentalis *Common name:* White cedar; *Local name:* Northern white cedar (USA), eastern arborvitae (USA), eastern white cedar (Canada) (not to be confused with *Cedrus*); *Source of origin:* Eastern north America; *Colour:* Lt. brown		7, 9, 10	16	21	26		30	34

Thuja plicata

Common name: Western red cedar: *Local names:* Giant arborvitae (USA), British Columbia red cedar (UK), red cedar (Canada) (not to be confused with *Cedrus*); *Sources of origin:* British Columbia, west USA, British Isles; *Colour:* Red-brown

Tsuga canadensis

Common names: Eastern or white hemlock; *Local name:* Hemlock spruce (USA); *Sources of origin:* Eastern Canada & USA; *Colour:* Lt. brown

Tsuga heterophylla

Common name: Western hemlock; *Local names:* Pacific hemlock (USA), British Columbian hemlock (USA), Alaska pine (Canada) (sometimes shipped with western balsam *(Abies)* and mountain hemlock (*T. mertensiana*); *Sources of origin:* British Columbia, Alaska, West USA, British Isles; *Colour:* Yellow-cream, or lt. brown

Tsuga mertensiana

Common name: Mountain hemlock (sometimes shipped with western hemlock, *T. heterophylla*); *Sources of origin:* West USA, Canada and Alaska

Hardwoods—Dicotyledonae

Acacia aneura

Common name: Mulga; *Source of origin:* Australia; *Colour:* Dk. brown and yellow; *Note:* very heavy

Acacia melanoxylon

Common name: Australian blackwood; *Local name:* Black wattle (Australia); *Source of origin:* Australia; *Colour:* Gold-dk. brown

Acacia mollissima

Common name: Black wattle; *Source of origin:* South Africa

Acanthopanax ricinifolius

Common name: Sen; *Source of origin:* Japan

Acer spp.—

Common name: Maple (English), Erable (French), Esdoorn (Dutch), Ahorn (German)

Acer campestre

Common name: Field maple; *Source of origin:* British Isles

Acer macrophyllum

Common name: Pacific maple; *Local names:* Oregon maple (USA), big leaf maple (Canada); *Sources of origin:* Pacific coast of USA and Canada; *Colour:* Pinkish-brown

Acer mono

Common name: Japanese maple (may include other species); *Source of origin:* Japan

Acer platanoides

Common name: Norway maple; *Local names:* European or Bosnian maple (UK); *Sources of origin:* Europe, British Isles

Acer pseudoplatanus

Common name: Sycamore; *Local names:* Plane (UK), sycamore plane (UK), great maple (UK) (not to be confused with USA sycamore, *Platanus occidentalis*); *Sources of origin:* Europe, British Isles; *Colour:* Whitish or red-brown

Species	*Uses*		*Properties*					
	Mode of use	*Typical uses*	*Texture*	*Density*	*Strength*	*Movement*	*Durability*	*Permeability*
Thuja plicata	A	2, 3, 10, 11	14	21	25	29	31	35
Tsuga canadensis		1, 4, 6, 11, 12	14	21	25		33	35
Tsuga heterophylla	A	1, 4, 5, 6, 11, 12	16	20	24	29	33	36
Tsuga mertensiana								
Acacia aneura	A	6		19			31	
Acacia melanoxylon	A	4, 6, 12	15, 18	20		28	31	
Acacia mollissima				19		27		
Acanthopanax ricinifolius								
Acer spp.								
Acer campestre				19				
Acer macrophyllum		6		20				
Acer mono				19				
Acer platanoides				19				
Acer pseudoplatanus	A	4, 6	16	20		28	33	37

	Uses		Properties					
	Mode of use	*Typical uses*	*Texture*	*Density*	*Strength*	*Movement*	*Durability*	*Permeability*
Acer rubrum *Common name:* Soft maple (includes *A. saccharinum*); *Local name:* Red maple (Canada, USA); *Sources of origin:* Canada, east USA; *Colour:* Lt. brown		6, 11		20	24		33	
Acer saccharinum *Common name:* Soft maple (includes *A. rubrum*); *Local name:* Silver maple (Canada, USA); *Sources of origin:* Canada, East USA				20				36
Acer saccharum (A. nigrum) *Common name:* Rock maple; *Local names:* Hard maple (UK, Canada, USA), white maple (sapwood only, USA); *Sources of origin:* Canada, east USA; *Colour:* White or lt. red-brown	A, I	5, 6, 13	16, 18	19		28	33	37
Achras zapota, see ***Manilkara zapota***								
Ackama australiensis *Common name:* Rose alder; *Local name:* Feathertop, pencil cedar (Australia); *Source of origin:* Australia; *Colour:* Lt. pinkish brown	A, I	4		19				
Ackama paniculata *Common name:* Brown alder; *Local names:* Corkwood, pencil cedar (Australia), Rose alder (including *A. australiensis); Source of origin:* Australia; *Colour:* Lt. pinkish brown	A, I,	4		19				
Adina cordifolia *Common name:* Haldu; *Local names:* Hnaw (Burma), kwao or kwow (Thailand); *Sources of origin:* Burma, India, Thailand				19			32	37
Aesculus hippocastanum *Common name:* European horse-chestnut; *Source of origin:* British Isles				20		29	33	37
Aesculus turbinata *Common name:* Japanese horse-chestnut; *Source of origin:* Japan				20				
Aextoxicon punctatum *Common name:* Olivillo; *Source of origin:* Chile				20		28		
Afrormosia elata, see ***Pericopsis elata***								
Afzelia africana *Common name:* Afzelia (includes other species); *Local names:* Doussié (Cameroons, France), apa (Nigeria), aligna (Nigeria) (these alternatives include *A. bipindensis* and *A. pachyloba*), lingué (Ivory Coast), papao (Ghana); *Source of origin:* West Africa; *Colour:* Lt. red-golden brown		2, 3, 4, 5, 6, 7, 8, 12	14, 17, 18	19	24	29	30	34
Afzelia bipindensis *Common name:* Afzelia (includes other species); *Local names:* Doussié (Cameroons, France), apa (Nigeria), aligna (Nigeria) (these alternatives include *A. africana* and *A. pachyloba*); *Source of origin:* West Africa; *Colour:* Lt. red-golden brown		3, 5, 6, 7, 8, 12	14, 17, 18	19	24	29	30	34
Afzelia pachyloba *Common name:* Afzelia (includes other species); *Local names:* Doussié (Cameroons, France), apa (Nigeria), aligna (Nigeria) (these alternatives include *A. africana* and *A. bipindensis*); *Source of origin:* West Africa; *Colour:* Lt. red-golden brown		3, 5, 6, 7, 8, 12	14, 17, 18	19	24	29	30	34
Afzelia quanzensis *Common name:* Afzelia (includes other species); *Local names:* Chamfuta (Mozambique), mussacossa (Mozambique), mkora (Tanzania), mbembakofi (Tanzania); *Source of origin:* East Africa				19			30	34

Species	Uses: Mode of use	Uses: Typical uses	Properties: Texture	Properties: Density	Properties: Strength	Properties: Movement	Properties: Durability	Properties: Permeability
Albizia adianthifolia *Common name:* West African albizia (includes other species); *Local names:* Ayinre (Nigeria), okuro (Ghana), sifou (France, Zaire) (these alternatives include other species); *Source of origin:* West Africa				20			30	34
Albizia ferruginea *Common name:* West African albizia (includes other species); *Local names:* Ayinre (Nigeria), okuro (Ghana), sifou (France, Zaire) (these alternatives include other species); *Source of origin:* West Africa				20			30	34
Albizia grandibracteata *Common name:* Nongo (may include *A. zygia); Local name:* Red or white nongo, according to colour; *Source of origin:* East Africa				20				
Albizia lebbek *Common name:* Kokko; *Local names:* East Indian walnut (UK), siris (India); *Sources of origin:* India, Andaman Islands, Burma				19			32	
Albizia toona *Common name:* Red siris; *Local name:* Acacia cedar (Australia); *Source of origin:* Australia; *Colour:* Dk. red	A, I	1, 6		19			32	
Albizia xanthoxylon *Common name:* Yellow siris; *Local name:* Yellow bean (sometimes confused with *Flindersia* spp.); *Source of origin:* Australia; *Colour:* Pale yellow		3, 4, 6, 7		20			32	
Albizia zygia *Common name:* West African albizia (includes other species); *Local names:* Ayinre (Nigeria), okuro (Ghana), sifou (France, Zaire), nongo (East Africa) (these alternatives include other species); *Sources of origin:* West Africa, east Africa				20				
Alnus **spp.—** *Common names:* Alder (English), Aune (French), Els (Dutch), Erle (German)								
Alnus glutinosa *Common name:* Common alder; *Local name:* Black alder (UK); *Sources of origin:* Europe and British Isles				20			33	37
Alnus incana *Common name:* Grey alder; *Source of origin:* North Europe				20				
Alstonia boonei, **see *A. congensis***								
Alstonia congensis *Common name:* Alstonia (includes *A. boonei); Local names:* Patternwood (E. and W. Africa), stoolwood (E. and W. Africa) mujua (Uganda), ahun or áwun (Nigeria), duku (Nigeria), tsongutti (Zaire), sindru (Ghana), emien (France, Ivory Coast); *Source of origin:* Tropical Africa				21		29	33	37
Alstonia scholaris *Common name:* White cheesewood; *Local names:* Milky pine, milkwood; *Source of origin:* Australia; *Colour:* Cream-yellow white				21			33	
Amblygonocarpus andogensis *Common name:* Banga wanga; *Source of origin:* East Africa; *Note:* Very heavy				19		29	31	
Amoora polystachys **and *A. cucullata*** *Common name:* Tasua; *Source of origin:* Thailand				20				35
Andira inermis *Common name:* Angelin; *Local names:* Angelim (Brazil), partridge wood (*UK), kuraru (Guyana), (red) cabbage-bark (British Honduras, USA) (*not to be confused with *Caesalpinia granadillo); Sources of origin:* West Indies, Central America, tropical S. America				19			31	

Species	Uses: Mode of use	Uses: Typical uses	Properties: Texture	Properties: Density	Properties: Strength	Properties: Movement	Properties: Durability	Properties: Permeability
Androstachys johnsonii *Common name:* Mecrusse; *Source of origin:* Mozambique; *Note:* Very heavy				19		28	31	
Aningeria altissima* and *A. robusta *Common name:* Aningeria; *Local names:* Aniègre (France, Ivory Coast), landosan (Nigeria), muna (Kenya), osan (Uganda), anegré (UK), Tanzanian walnut (UK); *Sources of origin:* West and East Africa				20				
Anisoptera costata *Common names:* Mersawa (Malaysia), Pengiran (Sabah), (these include *A. laevis and A. scaphula); Sources of origin:* Malaysia, Sabah				20		28	32	35
Anisoptera curtisii *Common name:* Krabak (includes *A. oblonga* and *A. scaphula*); *Source of origin:* Thailand				19		28	32	
Anisoptera laevis *Common names:* Mersawa (Malaysia), Pengiran (Sabah) (these include *A. costata* and *A. scaphula*); *Sources of origin:* Malaysia, Sabah				20		28	32	
Anisoptera oblonga *Common name:* Krabak (includes *A. curtisii* and *A. scaphula*); *Source of origin:* Thailand				20		28	32	
Anisoptera polyandra *Common name:* Anisoptera; *Local names:* Mersawa (Malaysia), Mentasawa, Garawa, Karawa; *Sources of origin:* Malaysia, Papua, New Guinea, Thailand, Indonesia, Burma, Philippines; *Colour:* Pale yellow-lt. brown	I	4, 6		20	25		33	
Anisoptera scaphula *Common name:* Kaunghmu (Burma); *Local names:* Mersawa (Malaysia) (includes *A. costata* and *A. laevis*), krabak (Thailand) (includes *A. curtisii* and *A. oblonga*); *Sources of origin:* Burma, Malaysia, Thailand				20		28	32	
Anisoptera thurifera *Common name:* Palosapis (may include other species); *Source of origin:* Philippines				19		28	32	
Anogeissus acuminata *Common names:* Yon (Burma), ta khian nu (Thailand); *Sources of origin:* Burma, India, Thailand; *Note:* Very heavy				19				35
Antiaris toxicaria (A. africana, A. welwitschii) *Common name:* Antiaris; *Local names:* Oro (Nigeria), ogiovu (Nigeria), kirundu (Uganda), kyenkyen or chenchen (Ghana), ako (France, Ivory Coast); *Source of origin:* Tropical Africa				21		29	33	37
Aspidosperma peroba (A. polyneuron) *Common name:* Peroba rosa; *Local name:* Red peroba (UK) (not to be confused with white peroba, *Paratecoma peroba)*; *Source of origin:* Brazil				19				
Astronium fraxinifolium* and *A. graveolens *Common name:* Goncalo alves; *Local names:* Zebrawood (UK), courbaril (UK), locustwood (UK), kingwood (USA), tigerwood (USA), urunday (Brazil); *Source of origin:* Brazil; *Note:* Very heavy				19				
Aucoumea klaineana *Common name:* Gaboon; *Local names:* Okoumé (Gabon), Gaboon mahogany (UK); *Sources of origin:* Gabon, Spanish Guinea				21			33	
Backhousia bancroftii *Common name:* Johnstone river hardwood; *Source of origin:* Australia; *Colour:* Lt. grey; *Note:* Very heavy		1, 5	16, 17	19				

	Uses		Properties					
	Mode of use	*Typical uses*	*Texture*	*Density*	*Strength*	*Movement*	*Durability*	*Permeability*
Baikiaea plurijuga *Common name:* Rhodesian teak; *Local names:* Zambesi redwood (Rhodesia), umgusi, mukushi or mukusi (Rhodesia); *Sources of origin:* Rhodesia, Zambia; *Note:* Very heavy				19		29	30	
Baillonella toxisperma (Minusops djave, M. toxisperma) *Common name:* Moabi (France and Gabon); *Local name:* Djave (Nigeria); *Source of origin:* West Africa				19				
Balfourodendron riedelianum *Common names:* Pau marfim (Brazil) (sometimes used for *Aspidosperma* spp.), (Guatambu) moroti (Argentina); *Source of origin:* Brazil, Argentina				19				
***Beilschmiedia* spp.** *Common name:* Brown walnut (Australia); *Source of origin:* Australia; *Colour:* Lt. brown	A, I	1, 4, 5, 6		19			33	
Beilschmiedia bancroftii *Common name:* Yellow walnut (Australia); *Local name:* Canary ash (Queensland); *Source of origin:* Australia; *Colour:* Pale yellow	I			20			33	
Beilschmiedia tawa *Common name:* Tawa; *Local name:* New Zealand oak (UK); *Source of origin:* New Zealand				19			33	35
Berlinia bracteosa, B. confusa, B. grandiflora *Common name:* Berlinia; *Local name:* Ekpogoi (Nigeria), abem (Cameroons), ebiara (France, Gabon), rose zebrano (UK); *Source of origin:* West Africa				19		28	33	35
***Betula* spp.—** *Common names:* Birch (English), Bouleau (French), Berk (Dutch), Birke (German)								
Betula alleghaniensis (B. lutea)* and *B. lenta *Common name:* Yellow birch; *Local names:* Canadian yellow birch (UK), Quebec birch (UK), American birch* (UK), hard birch (Canada) (*includes *B. papyrifera*); *Sources of origin:* Canada, East USA; *Colour:* Brown	A, I	4, 6, 11, 13	18	19	23	27	33	36
Betula maximowieziana *Common name:* Japanese birch; *Source of origin:* Japan				19				
Betula papyrifera *Common name:* Paper birch; *Local names:* American birch* (UK), white birch (Canada) (*includes *B. alleghaniensis*); *Source of origin:* Canada, East USA; *Colour:* Brown	A, I	4, 6, 11, 13	18	20	25		33	36
Betula papyrifera,* var. *occidentalis *Common name:* Western paper birch; *Local names:* Western birch (Canada, USA), western white birch (Canada, USA); *Sources of origin:* Western Canada and USA; *Colour:* Brown	A, I	4, 6, 11, 13	18	20				
Betula pendula, B. pubescens (B. odorata), B. alba *Common name:* European birch; *Local names:* English, Finnish, Swedish birch, according to origin, silver birch (UK), white birch (UK); *Source of origin:* Europe including British Isles; *Colour:* Cream, lt. brown	I	5, 6, 11, 12	16	19		27	33	37
Bombax buonopozense *Common name:* West African bombax; *Local name:* Kapokier (France, Ivory Coast); *Source of origin:* West Africa; *Note:* Very light; *B. brevicuspe* and *B. chevalieri* are heavier and darker				21				
Bowdichia nitida *Common name:* Sucupira (applied also to *Diplotropis* spp. and *Ferreirea spectabilis*); *Local name:* Black sucupira (UK); *Source of origin:* Brazil; *Note:* Very heavy				19				

Species	Uses: Mode of use	Uses: Typical uses	Properties: Texture	Properties: Density	Properties: Strength	Properties: Movement	Properties: Durability	Properties: Permeability
Brachylaena hutchinsii *Common name:* Muhuhu; *Local name:* Muhugwe (Tanzania); *Source of origin:* East Africa; *Note:* Very heavy				19		29	30	34
Brachystegia boehmii *Common name: Mjombo; Source of origin:* Tanzania				19		27		34
Brachystegia eurycoma, B. leonensis, B. nigerica *Common name:* Okwen (Nigeria); *Local names:* Brachystegia (Nigeria), naga (Cameroons, France); *Source of origin:* West Africa				20		28	32	34
Brachystegia fleuryana *Common name:* Zebrano; *Source of origin:* West Africa				20				
Brachystegia spiciformis *Common name:* Mtondo; *Sources of origin:* Central and East Africa				19		28	33	34
Brosimum paraense *Common name:* Satiné; *Local name:* Muirapiranga (Brazil), satiné rubané (France); *Source of origin:* Tropical America; *Note:* Very heavy				19				
Brya ebenus *Common name:* Cocuswood; *Local names:* Granadillo; brown, green or Jamaica ebony (USA); *Sources of origin:* Jamaica, Cuba; *Note:* Very heavy				19				
Bulnesia arborea *Common name:* Verawood; *Local name:* Maracaibo lignum vitae (UK) (not to be confused with lignum vitae—*Guaiacum* spp.); *Source of origin:* Venezuela; *Note:* Very heavy				19				
***Buxus* spp.—** *Common names:* Box (English), Buis (French), Palmhout (Dutch), Buchsbaum (German)								
Buxus macowani *Common name:* East London boxwood; *Local name:* Cape box; *Source of origin:* South Africa; *Note:* Very heavy				19				
Buxus sempervirens *Common name:* European boxwood; *Local names:* Box (UK), Abassian, Iranian, Persian, Turkey boxwood according to origin; *Sources of origin:* Europe, British Isles, Mediterranean, W. Asia; *Note:* Very heavy				19				
Byrsonima coriacea,* var. *spicata *Common name:* Serrete; *Source of origin:* Trinidad				19		28	32	36
Caesalpinia echinata (Guilandina echinata) *Common name:* Brazilwood; *Local names:* Bahia wood (UK), Para wood (UK), Pernambuco wood (UK); *Source of origin:* Brazil: *Note:* Very heavy				19				
Caesalpinia granadillo *Common name:* Partridgewood; *Local names:* Maracaibo ebony (UK), granadillo* (Venezuela) (*not to be confused with *Brya ebenus); Source of origin:* Venezuela; *Note:* Very heavy				19				
Calophyllum brasiliense (C. calabra) *Common names:* Jacareuba (South America), Santa Maria (Central America); *Sources of origin:* Tropical central and South America				20		28	31	34
Calophyllum cunstleri, C. curtisii *Common name:* Bintangor; *Source of origin:* Malaysia								

	Uses		Properties					
	Mode of use	Typical uses	Texture	Density	Strength	Movement	Durability	Permeability
Calophyllum inophyllum, C. kajewski, C. costatum *Common name:* Calophyllum; *Local names:* Beech calophyllum *(C. inophyllum)*, bush calophyllum (*C. kajewski*), red touriga (*C. costatum*)*; *Sources of origin:* East Indies, Pacific Islands, *North Queensland; *Colour:* Deep red	A	4, 5, 6, 7		19			32	
Calophyllum tomentosum *Common name:* Poon; *Source of origin:* India				19				
Calycophyllum candidissimum *Common name:* Degame; *Local names:* Degame lancewood (UK), lemonwood (USA); *Sources of origin:* Tropical Central and South America, Cuba				19				
***Campnosperma* spp.** *Common name:* Terentang; *Source of origin:* Malaysia				21				36
***Canarium* spp.** *Common name:* Malaysian canarium; *Local name:* Kedondong (Malaysia); *Source of origin:* Malaysia				20				34
Canarium euphyllum *Common name:* Indian canarium; *Local names:* Indian white mahogany (UK), (white) dhup (India); *Source of origin:* Andaman Islands				21			33	34
Canarium schweinfurthii *Common name:* African canarium; *Local names:* Abel (Cameroons), aiélé (France and Ivory Coast), mwafu (Uganda), abé (Spanish Guinea); *Sources of origin:* West and East Africa				20		28	33	34
Carapa guianensis *Common name:* Andiroba; *Local names:* Crabwood (Guyana), Brazilian, British Guiana, Surinam, Demerara or Para mahogany (UK), figueroa (Ecuador), tangare (Ecuador), krappa (Surinam); *Sources of origin:* Tropical South America, West Indies				20		29	32	
Cardwellia sublimis *Common name:* Australian silky-oak; *Local name:* Northern silky-oak (Australia); *Source of origin:* Australia; *Colour:* Lt. pink-brown (silver figure radial)	A, I	6		20			32	
Cariniana* spp., including *C. legalis, C. pyriformis *Common name:* Jequitiba (Brazil); *Local names:* Jequitiba rosa (Brazil), bacu (Venezuela), abarco or albarco (Colombia); *Source of origin:* Tropical South America				20				
Carpinus spp.— *Common names:* Hornbeam (English), Charme (French), Haagbeuk (Dutch), Hainbuche (German)				19			32	35
Carpinus betulus *Common name:* Hornbeam; *Sources of origin:* Europe, British Isles; *Colour:* Whitish			16	19		28	33	37
Carya aquatica *Common name:* Pecan (includes *C. illinoensis); Local names:* Water hickory (USA), bitter pecan (USA); *Source of origin:* USA; *Colour:* Red-brown		5, 6	18	19			33	35
Carya glabra *Common name:* Hickory (includes other species); *Local name:* Pignut hickory (USA) (Canadian hickory, usually *C. glabra* and *C. ovata*); *Sources of origin:* Eastern Canada and USA; *Colour:* Lt. pink or brown		6	18	19				
Carya illinoensis (C. pecan) *Common name:* Pecan (includes *C. aquatica*); *Local names:* Pecan hickory (USA), sweet pecan (USA); *Source of origin:* USA; *Colour:* Red-brown		5, 6	18	19			33	35

Species	Uses: Mode of use	Uses: Typical uses	Properties: Texture	Properties: Density	Properties: Strength	Properties: Movement	Properties: Durability	Properties: Permeability
Carya laciniosa *Common name:* Hickory (includes other species); *Local name:* Shellbark hickory (USA); *Sources of origin:* Eastern Canada and USA; *Colour:* Lt. pink or brown		6	18	19				
Carya ovata *Common name:* Hickory (includes other species); *Local names:* Shagbark hickory (USA) (Canadian hickory, usually *C. glabra* and *C. ovata); Sources of origin:* Eastern Canada and USA; *Colour:* Lt. pink or brown		6	18	19	22		33	
Carya tomentosa (C. alba) *Common name:* Hickory (includes other species); *Local name:* Mockernut hickory (USA); *Sources of origin:* Eastern Canada and USA; *Colour:* Lt. pink		6	18	19				
Casearia **spp., see *Gossypiospermum***								
Cassipourea malosana (C. elliottii) *Common name:* Pillarwood; *Local names:* Ndiri (Tanzania), musaisi (Kenya); *Source of origin:* East Africa				19		28	33	34
Cassipourea verticillata *Common name:* Mzimbe; *Local name:* Onionwood; *Source of origin:* Mozambique				20			31	
Castanea **spp.—** *Common names:* Chestnut (English), Châtaignier (French), Kastanje (Dutch), Kastanie (German)								
Castanea dentata *Common name:* American chestnut; *Source of origin:* USA; *Colour:* Brown	U	4, 6, 9, 10 11, 13	18	20	25		31	
Castanea sativa *Common name:* Sweet chestnut; *Local names:* Spanish chestnut (UK), European chestnut (UK); *Sources of origin:* Europe, British Isles; *Colour:* Lt. brown		4	15	20		29	31	34
Castanospermum australe *Common name:* Black bean; *Source of origin:* Australia; *Colour:* Dk. brown	A, I	4, 6	15	19		28	32	
Casuarina fraseriana *Common name:* Western Australian sheoak; *Source of origin:* Australia; *Colour:* Brown		6		19	25		31	
Catostemma commune *Common name:* Baromalli; *Source of origin:* Guyana				20		27	33	37
Cedrela fissilis *Common name:* South American cedar (not to be confused with cedar-*Cedrus); Local names:* Brazilian, Peruvian, etc. cedar according to origin (UK), cedar (UK), cigar-box cedar (UK), cedro (S. America); *Source of origin:* Tropical South America; *Colour:* Red-brown		4, 6	14	21		29	32	34
Cedrela odorata (C. mexicana) *Common name:* Central American cedar (not to be confused with cedar-*Cedrus); Local names:* Honduras, Mexican, Nicaraguan, Trinidad, etc. cedar according to origin (UK), cedar (UK), cigar-box cedar (UK), Spanish cedar; *Sources of origin:* Central America, West Indies; *Colour:* Red-brown		4, 6	14	20		29	31	36
Cedrela toona,* see *Toona ciliata								
Cedrela toona,* var. *australis,* see *Toona australis								
Ceiba occidentalis *Common name:* Ceiba (includes *C. pentandra*); *Local name:* Honduras cottonwood (UK); *Source of origin:* Tropical America; *Note:* Very light				21			33	

Species	Uses: Mode of use	Uses: Typical uses	Properties: Texture	Properties: Density	Properties: Strength	Properties: Movement	Properties: Durability	Properties: Permeability
Ceiba pentandra *Common name:* Ceiba (includes *C. occidentalis); Local names:* Silk cotton (Africa), fromager (France), fuma (Congo); *Sources of origin:* African and Asian tropical zones; *Note:* Very light				21			33	
Celtis occidentalis *Common name:* Hackberry; *Source of origin:* USA; *Colour:* Yellowish to lt. brown		6, 11		20	25		33	
Celtis* spp. *(C. adolfi-friderici, C. mildbraedii, C. zenkeri (C. soyanxii)) *Common name:* African celtis; *Local names:* Esa (Ghana), ita (Nigeria), ohia (Nigeria); *Source of origin:* Tropical Africa				19		28	33	36
Cephalosphaera usambarensis *Common name:* Mtambara; *Source of origin:* Tanzania				20				
Ceratopetalum apetalum *Common name:* Coachwood; *Source of origin:* Australia; *Colour:* Pink-brown	A, I	4, 6	16	20			33	37
Cercidiphyllum japonicum *Common name:* Katsura; *Source of origin:* Japan				21		29		
Chlorophora excelsa* and *C. regia *Common name:* Iroko; *Local names:* Mvule (East Africa), odum (Ghana, Ivory Coast), kambala (Zaire), tule or intule (Mozambique), moreira (Angola); *Sources of origin:* West and East Africa (*C. regia*—W. Africa only); *Colour:* Golden brown		1, 3, 4, 5, 6, 7, 8, 12, 13	14, 17	19	24, 25	29	30	34
Chlorophora tinctoria *Common name:* Fustic: *Local names:* Moralfino (Ecuador), yellowwood (West Indies), tatajuba (Brazil); *Sources of origin:* West Indies, tropical America				19				
Chloroxylon swietenia *Common name:* Ceylon satinwood; *Local name:* East Indian satinwood (UK); *Sources of origin:* Sri Lanka, India; *Note:* Very heavy				19			31	34
Chrysophyllum cainito *Common name:* Star apple; *Local name:* Longui rouge; *Source of origin:* Tropical America				19		28	31	
Chukrasia tabularis (C. velutina) *Common name:* Chickrassy; *Local names:* Chittagong wood (UK), yinma (Burma), yom hin (Thailand); *Sources of origin:* Burma, India, Pakistan, West Malaysia, Thailand				20			33	34
Cinnamomum lautrattii *Common name:* Pepperwood; *Source of origin:* Australia; *Colour:* Cream	A, I	4		21			33	
***Cistanthera*, see *Nesogordonia* spp.**								
Cleistocalyx gustavioides *Common name:* Grey satinash (Australia); *Local name:* Water gum (Australia); *Source of origin:* Australia; *Colour:* Yellow-grey	A,	1, 4, 5, 6, 7	15, 17	19			32	
Cleistopholis patens *Common name:* Otu (Nigeria); *Local name:* Sobu (France, Ivory Coast); *Source of origin:* West Africa				21				
Combretodendron macrocarpum (C. africanum) *Common name:* Esia (Ghana); *Local names:* Owawe (Nigeria), minzu (Congo); *Source of origin:* West Africa				19		27	31	
***Copaifera* spp., see *Pseudosindora* spp.**								

Species	Uses: Mode of use	Uses: Typical uses	Properties: Texture	Properties: Density	Properties: Strength	Properties: Movement	Properties: Durability	Properties: Permeability
Cordia abyssinica *Common name:* African cordia (also includes *C. millenii* and *C. platythyrsa*); *Local names:* Mukumari (Kenya), mringaringa (Tanzania); *Source of origin:* East Africa				20				
Cordia alliodora *Common name:* American light cordia (also includes *C. trichotoma*); *Local names:* Salmwood (British Honduras), laùrel de costa (Ecuador), Ecuador laurel (UK); *Sources of origin:* West Indies, tropical America				20			32	
Cordia goeldiana *Common name:* Freijo; *Source of origin:* Brazil				20			31	
Cordia millenii *Common names:* African cordia (also includes *C. abyssinica* and *C. platythyrsa*); *Local names:* Mugoma (Kenya), mukebu (Uganda), ome* (Nigeria) (*also includes *C. platythyrsa*); *Sources of origin:* West and East Africa				20				
Cordia platythyrsa *Common names:* African cordia (also includes *C. abyssinica* and *C. millenii*); *Local names:* Omo (Nigeria) (also includes *C. millenii); Source of origin:* West Africa				20				
Cordia trichotoma *Common names:* American light cordia (also includes *C. alliodora*); *Local names:* Louro* (Brazil), peterebi (Argentina) (*used elsewhere for *Lauraceae*, e.g. *Ocotea* spp.); *Sources of origin:* West Indies, tropical America				20				
Cornus florida *Common name:* Dogwood; *Local name:* Cornel (USA); *Source of origin:* USA; *Colour:* Sapwood, lt. pinkish brown; heartwood darker				19	23		33	
Cratoxylon arborescens *Common name:* Geronggang; *Sources of origin:* Sarawak, Malaysia; *Colour:* Pink-red brown		4	14, 18	20	25		33	
Croton megalocarpus *Common name:* Musine; *Source of origin:* East Africa				19		27	33	
Cryptocarya erythroxylon *Common name:* Rose maple (Australia); *Source of origin:* Australia; *Colour:* Lt. pink-brown	A, I	5, 6		19				
Cryptocarya glaucescens *Common name:* Silver sycamore (Australia); *Local names:* Brown beech (Australia), native laurel (Australia), jackwood; *Source of origin:* Australia; *Colour:* Pale yellow-brown	A, I	4, 6, 11		20				
Cryptocarya oblata *Common name:* Bolly silkwood; *Local names:* Tarzali, targali silkwood; *Source of origin:* Australia; *Colour:* Pink-red brown	I	6	18	20			33	
Cybistax **spp., see *Tabebuia* spp.**								
Cylicodiscus gabunensis *Common name:* Okan (Nigeria); *Local name:* Denya (Ghana); *Source of origin:* West Africa; *Note:* Very heavy				19			30	34
Cynometra alexandri *Common name:* Muhimbi; *Local name:* Muhindi; *Source of origin:* Uganda; *Note:* Very heavy				19		28	31	
Cynometra webberi *Common name:* Mfunda						27		

Species	Uses		Properties					
	Mode of use	Typical uses	Texture	Density	Strength	Movement	Durability	Permeability
Dactylocladus stenostachys *Common name:* Jongkong; *Local names:* Merebong or merubong (Sarawak), medang jongkong (Sarawak), medang tabak (Sarawak, Sabah); *Source of origin:* Sarawak, Sabah				20				
***Dalbergia* spp.—** *Common names:* Rosewood (English), Palissandre (French), Palissander (Dutch), Rosenholz (German)								
Dalbergia cearensis (?) *Common name:* Kingwood: *Local names:* Violetta or violet wood (USA); *Source of origin:* Brazil; *Note:* Very heavy				19				
Dalbergia frutescens*, var. *tomentosa *Common name:* Brazilian tulip wood; *Local names:* Pinkwood (USA), bois de rose (France) (not to be confused with tulip tree—*Liriodendron tulipifera*); *Source of origin:* Brazil; *Note:* Very heavy				19				
Dalbergia latifolia *Common name:* Indian rosewood; *Local names:* Bombay blackwood (India), East Indian rosewood (UK); *Sources of origin:* India, Java; *Colour:* Purplish brown		4	15	19		29	30	
Dalbergia melanoxylon *Common name:* African blackwood; *Local names:* Mozambique ebony (UK), mpingo (Tanzania); *Source of origin:* East Africa; *Note:* Very heavy				19		28		
Dalbergia nigra *Common name:* Brazilian rosewood; *Local names:* Bahia rosewood (UK), Rio rosewood (UK), jacaranda* Brazil) (*not to be confused with *Jacaranda* spp.); *Source of origin:* Brazil; *Colour:* Purplish brown		4	15	19		29		
Dalbergia oliveri *Common name:* Burma tulip wood; *Local name:* Tamalan (Burma) (not to be confused with tulip tree, *Liriodendron tulipifera); Source of origin:* Burma; *Note:* Very heavy				19			30	
Dalbergia retusa *Common name:* Cocobolo; *Source of origin:* Central America; *Note:* Very heavy				19				
Dalbergia sissoo *Common name:* Sissoo; *Local name:* Ṣhisham (Pakistan); *Sources of origin:* India, Pakistan				19				34
Dalbergia stevensonii *Common name:* Honduras rosewood; *Source of origin:* British Honduras; *Note:* Very heavy				19			30	
Daniellia ogea* and *D. thurifera *Common name:* Ogea; *Local names:* Oziya (Nigeria), daniellia (Nigeria), incenso or insenso (Portuguese Guinea), faro (France, Ivory Coast); *Source of origin:* West Africa				20		28	33	36
***Dialyanthera* spp.** *Common name:* Light virola; *Local name:* Virola; *Source of origin:* Columbia				21				
Dicorynia guianensis (D. paraensis) *Common name:* Basralocus; *Local names:* Angelique (French Guiana), Guyana teak (UK); *Sources of origin:* Brazil, the Guianas				19				
***Diospyros* spp.—** *Common names:* Ebony (English), Ébène (French), Ebben (Dutch), Ebenholz (German)								

Species	Uses: Mode of use	Uses: Typical uses	Properties: Texture	Properties: Density	Properties: Strength	Properties: Movement	Properties: Durability	Properties: Permeability
Diospyros celebica *Common name:* Macassar ebony; *Source of origin:* Celebes; *Colour:* Black or dk. brown; *Note:* Very heavy				19			31	35
Diospyros crassiflora* and *D. piscatoria *Common Name:* African ebony; *Local names:* Cameroons, Kribi, Gaboon, etc. ebony, according to origin (UK); *Source of origin:* Tropical Africa; *Colour:* Black or dk. brown; *Note:* Very heavy				19				
Diospyros ebenum *Common name:* Ceylon ebony; *Local name:* East Indian ebony (UK); *Sources of origin:* Sri Lanka, India; *Colour:* Black or dk. brown; *Note:* Very heavy				19			30	35
Diospyros marmorata *Common name:* Andaman marblewood; *Local name:* Zebrawood (UK); *Source of origin:* Andaman Islands; *Note:* Very heavy				19			30	35
Diospyros virginiana *Common name:* Persimmon; *Source of origin:* USA; *Colour:* Sapwood, creamy-white; heartwood, brown-black; *Note:* Mainly sapwood used commercially		11		19	23	27	31, 33	35
Dipterocarpus acutangulus, D. caudiferus, D. confertus, D. grandiflorus, D. warburgii *Common name:* Sabah keruing; *Local names:* Keruing or kruen (Sabah), sabah gurjun (UK); *Source of origin:* Sabah; *Colour:* Red-brown		1, 2, 3, 4, 5, 8, 9, 13	14, 18	19		28	32	37
Dipterocarpus acutangulus, D. apterus, D. caudiferus, D. lowii, D. verrucosus *Common name:* Sarawak keruing; *Local names:* Keruing (Sarawak), Sarawak gurjun (UK); *Sources of origin:* Sarawak, Brunei; *Colour:* Red-brown		1, 2, 3, 4, 5, 8, 9, 13	14, 18	19		28	32	37
Dipterocarpus alatus, D. turbinatus *Common names:* Gurjun (Burma), yang (Thailand); *Local name:* Kanyin (Burma); *Sources of origin:* Burma, Thailand; *Colour:* Red-brown		1, 2, 3, 4, 5		19		27	32	36
Dipterocarpus alatus, D. costatus, D. dyeri, D. intricatus, D. obtusifolius *Common name:* Dau; *Sources of origin:* South Vietnam, Khymer Republic				19			32	36
Dipterocarpus cornutus, D. costulatus, D. crinitus, D. sublamellatus *Common name:* Malaysian keruing; *Local names:* Keruing (Malaysia), Malaysian gurjun (UK); *Source of origin:* West Malaysia; *Colour:* Red-brown		1, 2, 3, 4, 5, 8, 9, 13	14, 18	19		28	32	37
Dipterocarpus cornutus, D. gracilis *Common name:* Indonesian keruing; *Local names:* Keruing, keroeing or keruwing (Indonesia), Indonesian gurjun (UK); *Sources of origin:* Indonesia, Sumatra, Kalimantan; *Colour:* Red-brown		1, 2, 3, 4, 5, 8, 9, 13	14, 18	19		28	32	37
Dipterocarpus gracilis, D. grandiflorus, D. lasiopodus *Common name:* Apitong; *Local names:* Bagac, Philippine gurjun (UK); *Source of origin:* Philippines; *Colour:* Red-brown		1, 2, 3, 4, 5, 8, 9, 13	14, 18	19			32	37
Dipterocarpus grandiflorus, D. indicus, D. macrocarpus *Common names:* Indian or Andaman gurjun; *Local names:* Gurjun (India), hollong (India); *Sources of origin:* India, Andaman Islands				19		27	32	36
Dipterocarpus tuberculatus *Common name:* Eng; *Local names:* In (Burma), pluang (Thailand); *Sources of origin:* Burma, Thailand				19			32	37

Species	*Uses*		*Properties*					
	Mode of use	*Typical uses*	*Texture*	*Density*	*Strength*	*Movement*	*Durability*	*Permeability*
Dipterocarpus zeylanicus *Common name:* Hora; *Local name:* Ceylon gurjun (UK); *Sources of origin:* Sri Lanka				19			32	36
Distemonanthus benthamianus *Common name:* Ayan (Nigeria); *Local names:* Anyaran (Nigeria), distemonanthus (UK), Nigerian (yellow) satinwood (UK), bonsamdua (Ghana), movingui (France, Ivory Coast); *Source of origin:* West Africa				19			32	35
Doryphora sassafras, D. aromatica *Common name:* Sassafras (may include *Daphnandra* spp. and *Dryadodaphre* spp.), southern sassafras is *Antherosperma moschatum* (similar properties); *Source of origin:* Australia; *Colour:* Yellow-green		4, 6		20			33	
Dracontomelum dao *Common name:* Paldao; *Local name:* Dao (Philippines); *Source of origin:* Philippines				19				
Dracontomelum mangiferum *Common name:* New Guinea walnut; *Local names:* Pacific walnut (UK), Papuan walnut (UK), loup or lup (new Guinea), laup (New Britain), damoui (Papua); *Sources of origin:* New Guinea and neighbouring islands; *Colour:* Pale black		6		20			33	
Dryobalanops aromatica *Common name:* Malaysian kapur (also includes *D. oblongifolia); Local names:* Kapur (West Malaysia), Sarawak kapur (also includes *D. lanceolata*), Indonesian kapur (also includes *D. oblongifolia, D. fusca, D. beccarii, D. lanceolata*); *Sources of origin:* West Malaysia, Sarawak, Indonesia (Sumatra, Kalimantan); *Colour:* Brown	A	1, 3, 4, 5, 6, 7, 12	14, 18	19		28		
Dryobalanops beccarii *Common names:* Sabah kapur or kapor (includes *D. lanceolata),* Indonesian kapur or kapoer (includes *D. aromatica, D. oblongifolia, D. fusca, D. lanceolata); Sources of origin:* Sabah, Indonesia (Kalimantan); *Colour:* Brown	A	1, 3, 4, 5, 6, 7, 12	14, 18	19		28	30	34
Dryobalanops fusca *Common names:* Indonesian kapur or kapoer (also includes *D. aromatica, D. oblongifolia, D. beccarii, D. lanceolata*); *Source of origin:* Indonesia (Kalimantan); *Colour:* Brown	A	1, 3, 4, 5, 6, 7, 12	14, 18	19		28		
Dryobalanops lanceolata *Common names:* Sarawak kapur (includes *D. aromatica*), Sabah kapur or kapor (includes *D. beccarii*), Indonesian kapur or kapoer (includes *D. aromatica, D. oblongifolia, D. fusca, D. beccarii*); *Sources of origin:* Sarawak, Sabah, Indonesia (Kalimantan); *Colour:* Brown	A	1, 3, 4, 5, 6, 7, 12	14, 18	19		28	30	35
Dryobalanops oblongifolia *Common names:* Malaysian kapur (also includes *D. aromatica*); *Local names:* kelandan (West Malaysia), Indonesian kapur or kapoer (also includes *D. aromatica, D. beccarii, D. fusca, D. lanceolata*); *Sources of origin:* West Malaysia, Indonesia (Sumatra, Kalimantan); *Colour:* Brown	A	1, 3, 4, 5, 6, 7, 12	14, 18	19		28		
Dyera costulata *Common name:* Jelutong; *Sources of origin:* Malaysia, Indonesia; *Colour:* Yellowish		4	16	21		29	32	37
Dysoxylum fraseranum *Common name:* Rose mahogany (Australia); *Local name:* Rosewood (New South Wales); *Source of origin:* Australia; *Colour:* Red-brown	A	4, 5, 6	16	19				
Dysoxylum muelleri *Common name:* Miva mahogany (Australia); *Local name:* Pencil cedar (Queensland); *Source of origin:* Australia; *Colour:* Red		1, 3, 4, 6	18	20				

Species	Uses: Mode of use	Uses: Typical uses	Properties: Texture	Properties: Density	Properties: Strength	Properties: Movement	Properties: Durability	Properties: Permeability
Elaeocarpus grandis *Common names:* Silver quandong (northern quandong, *E. foreolatus*; quandong, *E. largiflorens*); *Source of origin:* Australia; *Colour:* White-brown	A	4, 5, 6		21			33	
Endiandra palmerstonii *Common name:* Queensland walnut; *Local names:* Australian walnut (Australia), walnut bean (Australia), oriental wood (USA); *Source of origin:* Australia				19			33	
Endospermum medullosum, E. formicarium *Common name:* Basswood (Australia); *Source of origin:* Papua, New Guinea; *Colour:* White	A	4, 6, 11		21			33	
Entandrophragma angolense *Common name:* Gedu nohor (Nigeria); *Local names:* Edinam (Ghana), tiama (France, Ivory Coast) kalungi (Zaire); *Sources of origin:* West and East Africa; *Colour:* Lt. red-brown; *Note:* Variable durability	I	4, 5, 6, 7, 12	15, 17, 18	20	25	29	33	34
Entandrophragma candollei *Common name:* Omu (Nigeria); *Local names:* Heavy sapele (Nigeria), kosipo (France, Ivory Coast); *Source of origin:* West Africa; *Colour:* Dk. red-brown		5, 6, 12	15, 17	20	25	28	32	35
Entandrophragma cylindricum *Common name:* Sapele (Nigeria); *Local names:* Aboudikro (Ivory Coast), sapelli (France, Cameroons), cedar* (Ghana), sapele mahogany* (*not to be confused with true cedar or mahogany); *Sources of origin:* West and East Africa; *Colour:* Pink-red-brown	A, I	2, 3, 4, 5, 6, 12	15, 17	20	25	28	32	
Entandrophragma utile *Common name:* Utile (Ghana); *Local names:* Sipo (France, Ivory Coast), assié (Cameroons), cedar* (Ghana), utile mahogany* (* not to be confused with true cedar and mahogany); *Sources of origin:* West and East Africa	A, I	2, 3, 4, 5.	15	19		28	31	34
Eperua falcata *Common names:* Wallaba (includes *E. grandiflora*); *Local name:* Soft wallaba (Guyana); *Source of origin:* Tropical South America; *Note:* Very heavy				19			31	34
Eperua grandiflora *Common names:* Wallaba (includes *E. falcata*); *Local name:* Ituri wallaba (Guyana); *Source of origin:* Tropical South America; *Note:* Very heavy				19			31	34
Erythrophleum ivorense (E. micranthum) *Common names:* Missanda (Mozambique), kassa (Zaire), muave (Zambia, Mozambique); *Local names:* Tali (France, Ivory Coast), potrodom (Ghana), sasswood (Nigeria), erun (Nigeria); *Source of origin:* Tropical Africa; *Note:* Very heavy				19			30	34
Erythrophleum suaveolens (E. guineese) *Common name:* Mumara (Uganda); *Local names:* Tali (France, Ivory Coast), potrodom (Ghana), sasswood (Nigeria), erun (Nigeria) (these alternatives also include *E. ivorense*); *Source of origin:* Tropical Africa; *Note:* Very heavy				19		29	30	34
Eucalyptus acmenioides, E. umbra (E. carnea) *Common name:* Yellow stringybark; *Local name:* White mahogany (Australia); *Source of origin:* Australia: *Colour:* Yellow-brown; *Note:* Very heavy	A	8, 9, 12, 13	17	19	23		31	34
Eucalyptus astringens *Common name:* Brown mallet; *Source of origin:* Australia; *Colour:* Lt. red-brown; *Note:* Very heavy				19	23		32	
Eucalyptus bosistoana *Common name:* Coast grey box (Australia); *Source of origin:* Australia; *Colour:* Lt. brown; *Note:* Very heavy	A	8, 9, 13	17	19	22		30	

Species	Uses		Properties					
	Mode of use	Typical uses	Texture	Density	Strength	Movement	Durability	Permeability
Eucalyptus botryoides *Common name:* Southern mahogany (Australia); *Local names:* Bangalay, gippsland mahogany (Australia); *Source of origin:* Australia: *Colour:* Red-brown; *Note:* Very heavy	A	1, 13	14, 17	19	23		31	
Eucalyptus calophylla *Common name:* Marri; *Local name:* Red gum; *Source of origin:* Australia; *Colour:* Cream				19			32	
Eucalyptus camaldulensis *Common name:* River red gum; *Local names:* Red gum, river gum, Murray red gum; *Source of origin:* Australia	A	1, 8	16, 17	19	24		31	
Eucalyptus citriodora *Common names:* Spotted gum, macula (includes *E. maculata); Local name:* Lemon scented gum (Australia); *Source of origin:* Australia; *Note:* Very heavy	A	1, 2, 5, 7, 8, 11, 12, 13		19	22		32	34
Eucalyptus cloeziana *Common name:* Gympie messmate; *Local name:* Yellow messmate; *Source of origin:* Australia; *Colour:* Pale yellow; Note: Very heavy		1, 8, 10, 13		19	23		30	
Eucalyptus consideniana *Common name:* Yertchuk; *Local names:* Messmate, yellow messmate; *Source of origin:* Australia; *Colour:* Lt. brown; *Note:* Very heavy	A	1, 5, 13	14, 17	19	23		31	
Eucalyptus crebra *Common name:* Ironbark (includes other species); *Local name:* Narrow leaved red ironbark (Australia); *Source of origin:* Australia; *Colour:* Dk. red; *Note:* Very heavy		8, 9, 13		19	22		30	34
Eucalyptus cypellocarpa *Common name:* Mountain (grey) gum; *Source of origin:* Australia; *Colour:* Pale yellow-brown	A	1, 5, 8	16	19	24		32	
Eucalyptus deglupta *Common name:* Kamarere; *Sources of origin:* New Britain, Philippines; *Colour:* Red-brown	A	1, 2, 4, 7		20			33	
Eucalyptus delegatensis (E. gigantea) *Common name:* Tasmanian oak (includes other species); *Local names:* Alpine ash (Australia), woolly butt, red mountain ash, white or gum top stringybark; *Source of origin:* Australia; *Colour:* Pale brown	A	1, 2, 4, 5, 6	18	20	24	28	33	
Eucalyptus diversicolor *Common name:* Karri; *Source of origin:* Western Australia: *Colour:* Red-brown; *Note:* Very heavy	A, I	1, 4, 5, 6, 7, 8, 12		19	23	27	32	34
Eucalyptus drepanophylla, E. paniculata, E. siderophloia *Common name:* Ironbark (includes other species); *Local name:* Grey ironbark (Australia); *Source of origin:* Australia; *Colour:* Dark reddish brown; *Note:* Very heavy	A	8, 9,	17	19	22		30	34
Eucalyptus engenioides, E. phaeotricha, E. globoidea *Common name:* White stringybark; *Local name:* Pink blackbutt; *Source of origin:* Australia; *Colour:* Pale pink	A	1, 8, 9, 13	18	19	23		32	
Eucalyptus fastigata *Common name:* Brown barrel; *Local names:* Cut-tail, black mountain ash, silver or white top woolly butt; *Source of origin:* Australia; *Colour:* Pale brown	A	1, 5, 11	14, 18	19				
Eucalyptus fraxinoides *Common name:* Australian white ash; *Local name:* White ash (Australia); *Source of origin:* Australia				19			31	

Species	Uses		Properties					
	Mode of use	*Typical uses*	*Texture*	*Density*	*Strength*	*Movement*	*Durability*	*Permeability*
Eucalyptus globulus *Common name:* Southern blue gum (includes *E. stjohnii*); *Local name:* Tasmanium blue gum (Australia); *Source of origin:* Australia; *Colour:* Lt. brown	A	1, 8, 12	14	19	23		32	35
Eucalyptus gomphocephala *Common name:* Tuart; *Source of origin:* Australia; *Colour:* Lt. yellow; *Note:* Very heavy		7, 12	17	19	22		30	
Eucalyptus grandis *Common name:* Saligna gum (includes *E. saligna*); *Local names:* Rose gum (Australia), flooded gum (Australia); *Sources of origin:* Australia, South Africa; *Colour:* Dk. red; *Note:* South African wood is lower density		1, 2, 5, 11		19, 20				
Eucalyptus gummifera, E. intermedia *Common name:* Red bloodwood; *Local name:* Bloodwood; *Source of origin:* Australia; *Colour:* Dk. red	A	3, 10, 13	14	19	23		30	
Eucalyptus macrorrhyncha *Common name:* Red stringybark; *Local name:* Mountain stringybark; *Source of origin:* Australia; *Colour:* Pale red		1, 2, 9, 10	15	19				
Eucalyptus maculata *Common names:* Spotted gum, macula (includes *E. citriodora*); *Local name:* Maculata gum (South Africa); *Source of origin:* Australia; *Note:* Very heavy	A	1, 2, 5, 7, 8, 11, 12, 13	14, 17 18	19	23		32	34
Eucalyptus marginata *Common name:* Jarrah; *Source of origin:* Western Australia; *Colour:* Red-brown		1, 2, 3, 4, 5, 8, 9, 13	15	19	24	28	31	34
Eucalyptus melliodora *Common name:* Yellow box (Australia); *Source of origin:* Australia; *Colour:* Yellow-brown; *Note:* Very heavy		8, 9, 13		19	23		31	
Eucalyptus microcarpa, E. moluccana *Common name:* Grey box (Australia); *Local name:* Gum top box; *Source of origin:* Australia; *Colour:* Cream; *Note:* Very heavy		1		19	22		30	
Eucalyptus microcorys *Common name:* Tallowwood; *Source of origin:* Australia; *Colour:* Lt. yellow-brown; *Note:* Very heavy	A	1, 2, 3, 5, 8, 9, 12, 13		19	23	28	30	34
Eucalyptus obliqua *Common name:* Tasmanian oak (includes other species); *Local names:* Messmate stringybark (Australia), messmate, stringybark, brown-top stringybark; *Source of origin:* Australia; *Colour:* Lt. brown		1, 2, 4, 5, 6, 8, 10, 11, 13	14	19	23	28	32	35
Eucalyptus patens *Common name:* Western Australian blackbutt; *Source of origin:* Australia		1, 3, 4, 12, 13		19	23		32	
Eucalyptus pilularis *Common name:* Blackbutt; *Source of origin:* Australia; *Colour:* Brown; *Note:* Very heavy		1, 2, 5, 8, 13	14, 18	19	23		30	34
Eucalyptus polyanthemos *Common name:* Red box (Australia); *Source of origin:* Australia: *Colour:* Red; *Note:* Very heavy		8, 9, 13	17	19	23		31	
Eucalyptus regnans *Common name:* Tasmanian oak (includes other species); *Local names:* Mountain ash (Australia), Victorian ash (Australia), white ash (Australia), swamp gum; *Source of origin:* Australia; *Colour:* Lt. brown	A	1, 2, 4, 5, 6, 11	14, 18	19	24	28	33	

Species	Uses: Mode of use	Uses: Typical uses	Properties: Texture	Density	Strength	Movement	Durability	Permeability
Eucalyptus resinifera, E. pelleta *Common name:* Red mahogany (Australia); *Local names:* Red stringybark, red messmate; *Source of origin:* Australia; *Colour:* Dk. red; *Note:* Very heavy	A	1, 2, 5, 7, 12, 13	17	19	23		32	
Eucalyptus saligna *Common names:* Saligna gum (includes *E. grandis); Local names:* (Sydney) blue gum (Australia); *Sources of origin:* Australia, South Africa; *Colour:* Lt. red; *Note:* Australian wood very heavy, South African heavy	A	1, 2, 5, 13	14, 18	19	23		32	34
Eucalyptus sideroxylon *Common name:* Ironbark (includes other species); *Local name:* Red ironbark (Australia); *Source of origin:* Australia; *Colour:* Dk. red; *Note:* Very heavy		8, 9, 13		19	22		30	
Eucalyptus sieberi *Common name:* Silver-top ash (Australia); *Local names:* Coast or black or mountain ash (Australia), ironbark (Tasmania); *Source of origin:* Australia; *Colour:* Brown-pink		1, 5, 6, 12		19	23		33	
Eucalyptus stjohnii (E. bicostata) *Common names:* Southern blue gum (includes *E. globulus*); *Source of origin:* Australia				19			31	35
Eucalyptus tereticornis *Common name:* Forest red gum; *Local name:* Blue gum; *Source of origin:* Australia; *Colour:* Dk. red; *Note:* Very heavy		3, 5, 9, 10		19	23		31	
Eucalyptus torelliana *Common name:* Cadaga; *Source of origin:* Australia; *Colour:* Lt. brown; *Note:* Very heavy		1		19			32	
Eucalyptus viminalis *Common name:* Manna gum; *Source of origin:* Australia; *Colour:* Pale yellow-pink	A	1, 4, 5, 11, 12		19	24		33	
Eucalyptus wandoo (E. redunca, **var.** ***elata)*** *Common name:* Wandoo; *Source of origin:* Western Australia; *Note:* Very heavy				19			30	34
Eucryphia cordifolia *Common name:* Ulmo; *Source of origin:* Chile				20			33	
Euroschinus falcatus *Common name:* Pink poplar (Australia); *Local names:* Maiden's blush, blush cudgerie; *Source of origin:* Australia; *Colour:* Pale pink	A	6, 11	14, 18	20			33	
Eusideroxylon zwageri *Common name:* Belian, billian (Sabah); *Local name:* Borneo ironwood (UK); *Sources of origin:* Sabah, Sarawak, Indonesia; *Note:* Very heavy				19			30	34
Fagara flava (Zanthoxylum flavum) *Common name:* West Indian satinwood; *Local names:* Jamaica satinwood (UK), San Domingan satinwood (USA); *Source of origin:* West Indies; *Note:* Very heavy				19			33	35
Fagara heitzii *Common name:* Olon (France, Gabon); *Local name:* Olon tendre (France); *Source of origin:* West Africa				20				
Fagara macrophylla *Common name:* African satinwood; *Local names:* Olonvogo (France, Gabon), olon dur (France); *Sources of origin:* West and East Africa				19		28		

Species	Uses		Properties					
	Mode of use	Typical uses	Texture	Density	Strength	Movement	Durability	Permeability
Fagaropsis angolensis *Common name:* Mafu; *Local names:* Mfu (Tanzania), muruma (Kenya), mukarakati (Kenya); *Source of origin:* East Africa				19				
Fagus spp.— *Common names:* Beech (English), Hêtre (French), Beuk (Dutch), Buche (German)								
Fagus crenata *Common name:* Japanese beech (may include other species); *Local name:* Buna (Japan); *Source of origin:* Japan				20				
Fagus grandifolia *Common name:* American beech; *Sources of origin:* Canada, East USA; *Colour:* Whitish pink	A, I	5, 6, 11, 13		19	23		33	37
Fagus orientalis *Common name:* Turkish beech; *Sources of origin:* Turkey, neighbouring countries				19				
Fagus sylvatica *Common name:* European beech; *Local names:* English, Carpathian, Danish, French, Slavonian, etc. beech according to origin; *Sources of origin:* Europe, including British Isles; *Colour:* Pinkish red or lt. brown		4, 5, 6,	16, 18	19		27	33	37
Flindersia spp.— *Common name:* Silver ash; *Local names:* Queensland silver ash (*F. bourjotiana*); northern silver ash (*F. pubescens*), southern silver ash (*F. schottiana*), (*F. schottiana* also known as bumpy ash (Queensland) and cudgerie (New South Wales); *Source of origin:* Australia; *Colour:* White-light brown	A	6	17, 18	19			32	
Flindersia acuminata *Common name:* Silver silkwood; *Local name:* White silkwood, silver maple (Australia), Putt's pine (Australia); *Source of origin:* Australia; *Colour:* Pale yellow-brown	A	1, 6	18	20				
Flindersia australis *Common name:* Crow's ash; *Local name:* Australian teak; *Source of origin:* Australia; *Colour:* Yellow	A	3, 5, 7, 12, 13		19	23		30	
Flindersia brayleyana, F. pimenteliana *Common name:* Queensland maple; *Local names:* Australian maple (UK), maple silkwood (Australia); *Source of origin:* Australia; *Colour:* Lt. pinkish brown	A, I	4, 6, 7	17	20				
Flindersia ifflaiana *Common name:* Hickory ash; *Local names:* Cairn's hickory, hickory (Australia); *Source of origin:* Australia; *Colour:* Yellow-brown; *Note:* Very heavy	A	1, 5, 8,		19	23		30	
Flindersia laevicarpa *Common name:* Scented maple (Australia); *Local name:* Rose ash; *Source of origin:* Australia; *Colour:* Pale pink	A	1, 2, 3, 5	16	19			32	
Flindersia xanthoxyla *Common name:* Yellow-wood; *Local name:* Long jack; *Source of origin:* Australia; *Colour:* Yellow-brown	A	5, 7, 12	16, 17	19			32	
Fraxinus spp.— *Common names:* Ash (English), Frêne (French), Es (Dutch), Esche (German)								
Fraxinus americana *Common names:* American or Canadian ash (includes *F. nigra, F. pennsylvanica*); *Local name:* White ash (Canada); *Sources of origin:* Canada, USA; *Colour:* Lt. brown		6, 7, 11, 12	18	19	23		31	

Species	Uses: Mode of use	Uses: Typical uses	Properties: Texture	Properties: Density	Properties: Strength	Properties: Movement	Properties: Durability	Properties: Permeability
Fraxinus excelsior *Common name:* European ash; *Local names:* English, French, Polish, Slavonian, etc. ash, according to origin; *Sources of origin:* Europe, British Isles; *Colour:* Very lt. pinkish yellow-brown	A, I	4, 12	15	19		28	33	36
Fraxinus mandshurica *Common name:* Japanese ash; *Local name:* Tamo (Japan); *Source of origin:* Japan				19		27		
Fraxinus nigra *Common names:* American or Canadian ash (includes *F. americana, F. pennsylvanica); Local names:* Black ash (USA), brown ash (USA); Sources of origin: Canada, USA; *Colour:* Brown		6	18	20	25		31, 33	
Fraxinus pennsylvanica *Common names:* American and Canadian ash (includes *F. nigra, F. americana*); *Local names:* Green ash (USA), red ash (Canada); *Sources of origin:* Canada, USA; *Colour:* Lt. brown		6, 7, 11, 12	18	19				
Galbulimima baccata *Common name:* Magnolia (Australia); *Source of origin:* Australia; *Colour:* White to dk. brown	A, I	1, 4, 5		20			33	
Geissois benthami *Common name:* Brush mahogany (Australia); *Local names:* Red carabeen, red bean; *Source of origin:* Australia; *Colour:* Pink-brown	A, I	1, 4, 6, 11	16	20				
Gonystylus macrophyllum (bancanus, warburgianus) *Common name:* Ramin; *Local names:* Ramin telur (Sarawak), melawis (W. Malaysia) ahmin; *Sources of origin:* Sarawak, W. Malaysia; *Colour:* White-cream, or yellowish	A, I	4, 5, 6, 11	16, 17	19		27	33	
Gossweilerodendron balsamiferum *Common name:* Agba (Nigeria); *Local names:* Tola (France, Zaire), tola branca or white tola (Angola), (not to be confused with tchitola—*Oxystigma* spp.); *Source of origin:* West Africa; *Colour:* Yellow-lt. pink-brown	A, I	1, 3, 4, 5, 6, 7, 12	15, 16, 17	20	25	29	31	35
Gossypiospermum praecox (Casearia praecox) *Common name:* Maracaibo boxwood; *Local names:* Venezuelan, West Indian* or Colombian boxwood (UK), zapatero (*shipped via Curacao but not grown in West Indies); *Sources of origin:* Venezuela, Columbia				19				
Grevillea robusta *Common name:* Grevillea; *Local names:* African silky-oak (UK), southern silky-oak (Australia) (not be be confused with Australian silky-oak, *Cardwellia sublimis); Sources of origin:* Australia, East Africa				20			32	
Guaiacum* spp. (includes *G. officinale) *Common name:* Lignum vitae; *Local names:* Mexican, Nicaraguan lignum vitae (mixed species), Cuban, Jamaican, San Domingan, Puerto Rican lignum vitae (principally *G. officinale,* or thin-sap lignum vitae), Bahamas lignum vitae (principally *G. sanctum,* or thick-sap lignum vitae); *Sources of origin:* West Indies, tropical America; *Note:* Very heavy				19			30	34
Guarea cedrata *Common names:* Guarea (includes *G. thompsonii); Local names:* Pearwood (Nigeria), obobo (Nigeria), bossé (France, Ivory Coast) (these alternatives include *G. thompsonii),*white or scented guarea (Nigeria), obobonufua (Nigeria); *Source of origin:* West Africa; *Colour:* Pink-brown	A, I	2, 3, 4, 5, 6, 7, 12	15, 16, 18	20	24, 25	29	31	34
Guarea thompsonii *Common names:* Guarea (includes *G. cedrata); Local names:* Pearwood (Nigeria), obobo (Nigeria), bossé (France, Ivory Coast)(these alternatives include *G. cedrata),* black guarea (Nigeria), obobonekwi (Nigeria); *Source of origin:* West Africa; *Colour:* Pink-brown	A, I	2, 3, 4, 5, 6, 7	15	20		29	30	34

Species	Uses: Mode of use	Uses: Typical uses	Properties: Texture	Properties: Density	Properties: Strength	Properties: Movement	Properties: Durability	Properties: Permeability
Guibourtia arnoldiana *Common name:* Mutenye (Zaire); *Local names:* Benge or libengi (Zaire), olive walnut (UK); *Source of origin:* West Africa				19		28		
Guibourtia coleosperma *Common name:* Rhodesian copalwood; *Local names:* Umtjibi (Rhodesia), muzaule (Zambia), muxibe or mussibi (Angola), Rhodesian mahogany (Rhodesia); *Source of origin:* South central Africa				19				
Guibourtia demeusei, G. pellegrineana, G. tessmanii *Common name:* Bubinga (France, Cameroons); *Local names:* Kévazingo (Gabon), African rosewood (UK); *Source of origin:* West Africa				19				
Guibourtia ehie *Common name:* Ovangkol (France, Gabon); *Local names:* Amazakoué (Ivory Coast), anokye (Ghana), hyeduanini (Ghana); *Source of origin:* West Africa				19				
Haplormosia monophylla *Common name:* Haplormosia; *Local name:* Idewa (Gabon); *Source of origin:* West Africa				19				
Heritiera actinophylla *Common name:* Blush tulip oak (Australia); *Source of origin:* Australia; *Colour:* Brown	A, I	1, 4, 5, 6, 12	15, 17	19			33	
Heritiera cochinchinensis *Common name:* Chumprak; *Local name:* Chumprag (Thailand); *Source of origin:* Thailand				19				
Heritiera javanica and H. simplicifolia *Common name:* Mengkulang; *Local name:* Kembang (Sabah); *Source of origin:* Malaysia				19			33	35
Heritiera peralata *Common name:* Red tulip oak (Australia); *Source of origin:* Australia; *Colour:* Red-brown	A, I	1, 4, 5, 6		19				
Heritiera trifoliolata *Common name:* Brown tulip oak (Australia); *Local names:* Crow's foot elm, boojong, stavewood; *Source of origin:* Australia: *Colour:* Dk. Brown	A, I	1, 4, 5, 6, 12	15, 17	19			33	
***Hicoria* spp., see *Carya* spp.** *Common name:* Hickory								
Homalium foetidum *Common name:* Malas; *Source of origin:* Papua, New Guinea; *Colour:* Red-brown	A	1, 7, 8		19	23		31	
Hopea mengarawan, H. nervosa, H. pubescens *Common name:* Merawan (Malaysia); *Local name:* Selangan (Sarawak, Sabah); *Source of origin:* Malaysia				19				
Hopea nutans, H. pentanervia *Common name:* Giam; *Source of origin:* Malaysia; *Note:* Very heavy				19				
Hopea odorata *Common name:* Thingan (Burma); *Local names:* Takien (Thailand), sao (Vietnam); *Sources of origin:* India, Andaman Islands, Burma, Thailand, Khymer Republic, Vietnam				19			30	
Hura crepitans *Common name:* Hura; *Local names:* Sandbox, assacu (Brazil), possentrie (Surinam); *Sources of origin:* West Indies, tropical America				21			32	37

Species	Uses		Properties					
	Mode of use	Typical uses	Texture	Density	Strength	Movement	Durability	Permeability
Hymenaea courbaril *Common name:* Courbaril; *Local names:* Locust (Guyana), West Indian locust (UK); *Sources of origin:* West Indies, tropical America; *Note:* Very heavy				19			30	
Ilex aquifolium *Common name:* Holly; *Source of origin:* British Isles; *Colour:* White				19		27	33	
Intsia bijuga *Common names:* Merbau (may include *I. palembanica*); *Local names:* Hintzy (Madagascar), ipil (Philippines), kwila (New Guinea, New Britain), melila, bendora; *Sources of origin:* Madagascar, Philippines, South-west Pacific Islands; *Colour:* Dk. brown; *Note:* Very heavy	A	5, 7, 8		19	23		31	34
Intsia palembanica *Common name:* Merbau (may include *I. bijuga); Local name:* Borneo teak (UK); *Sources of origin:* Malaysia, Indonesia				19			31	34
Jacaranda copaia *Common names:* Futui or futi (Guyana); *Local name:* Parapara (Brazil); *Source of origin:* Tropical America				21				
***Juglans* spp.—** *Common names:* Walnut (English), Noyer (French), Walnoot (Dutch), Walnuß (German)								
Juglans nigra *Common name:* American walnut; *Local names:* Black walnut (USA, UK), walnut (USA); *Source of origin:* Eastern USA: *Colour:* Lt. or dk. brown	I	6, 10, 13	16, 18	19	23		32	
Juglans regia *Common name:* European walnut; *Local names:* English, French, Italian, Black Sea, Circassian, Persian walnut according to origin; *Source of origin:* Europe, British Isles, South-west Asia; *Colour:* Dk. brown		4, 6	16	19		28	32	35
Juglans sieboldiana *Common name:* Japanese walnut; *Local name:* Japanese claro walnut; *Source of origin:* Japan				21				
Kalopanax pictus*, see *Acanthopanax ricinifolius								
***Khaya* spp.—** *Common names;* African mahogany (English), Acajou d'Afrique (French)								
Khaya anthotheca *Common names:* African mahogany (includes *K. ivorensis, K. nyasica); Local names:* Ghana, Ivory Coast, Grand Bassam, etc. mahogany according to origin (includes *K. ivorensis*) Krala (Ivory Coast), mangona (Cameroons), munyama (Uganda); *Sources of origin:* West and East Africa; *Colour:* Lt. or dk. red-brown	A, I	2, 3, 4, 5, 6, 7, 12	15	20			32	34
Khaya grandifoliola *Common names:* Heavy African mahogany (includes *K. senegalensis*); *Local names:* Benin (Nigeria), grandifoliola (UK); *Source of origin:* West Africa	A, I	3, 4, 5, 6, 7, 12		19		28		34
Khaya ivorensis *Local names:* African mahogany (includes *K. anthotheca, K. nyasica*); *Local names:* Ghana, Ivory Coast, Grand Bassam, etc. mahogany, according to origin (includes *K. anthotheca),* Lagoswood (Nigeria), ogwango (Nigeria), ngollon (Cameroons); *Source of origin:* West Africa; *Colour:* Lt., dk. red-brown	A, I	2, 3, 4, 5, 6, 7, 12	15, 17, 18	20	25	29	32	34
Khaya nyasica *Common names:* African mahogany (includes *K. ivorensis, K. anthotheca*); *Local names:* Mozambique mahogany (UK), mbaua or umbaua (Mozabique), mbawa (Malawi), mkangazi (Tanzania); *Source of origin:* East Africa	A, I	2, 3, 4, 5, 6, 7, 12	15	20		29	32	34

	Uses		Properties					
	Mode of use	Typical uses	Texture	Density	Strength	Movement	Durability	Permeability
Khaya senegalensis *Common names:* Heavy African mahogany (includes *K. grandifoliola*); *Local names:* Bisselon (Port Guinea), Guinea mahogany (UK), dry-zone mahogany; *Sources of origin:* West and Central Africa				19		29		
Koompassia malaccensis *Common name:* Kempas; *Local name:* Impas (Sabah); *Source of origin:* Malaysia				19			31	37
Laburnum anagyroides (L. vulgare) *Common name:* Laburnum; *Source of origin:* British Isles; *Colour:* White				19				
Lagerstroemia hypoleuca *Common name:* Andaman pyinma; *Source of origin:* Andaman Islands				20				
Lagerstroemia speciosa (L. flos-reginae) *Common name:* Pyinma (Burma); *Local names:* Jarul (India, Pakistan), banglang (Vietnam), intanin (Thailand); *Sources of origin:* India, Burma, Thailand, Vietnam, West Malaysia				20			31	35
Laurelia philippiana (L. serrata) *Common name:* Tepa; *Source of origin:* Chile				20				
Laurelia sempervirens (L. aromatica) *Common name:* Chilean laurel; *Source of origin:* Chile				20				
Liquidambar styraciflua *Common names:* American red gum (heartwood), American sap gum (sapwood); *Local names:* (Sweet) gum (USA), satin walnut (heartwood, UK), hazel pine (sapwood, UK), bilstead (USA); *Source of origin:* South-eastern USA; *Colour:* White-pink sapwood, grey-brown heartwood		6, 11, 13	17	20				
Liriodendron tulipifera *Common name:* American whitewood; *Local names:* Canary (white) wood (UK), tulip (UK, Canada, USA), (yellow or tulip) poplar (USA) (not to be confused with whitewood—*Picea* and *Abies* spp.); *Sources of origin:* Canada and Eastern USA; *Colour:* White, yellowish or brownish	A, I	4, 6, 11	18	20				
Litsea reticulata *Common name:* Bollywood; *Local names:* Bolly gum, brown bollywood, brown beech (Australia); *Source of origin:* Australia; *Colour:* Pale brown-yellow-pink	A, I	4, 6	15	20			33	
Lophira alata *Common names:* Ekki, eba (Nigeria); *Local names:* Kaku (Ghana), azobé (France, Ivory Coast), bongossi (Cameroons); *Source of origin:* West Africa; *Colour:* Dk. red-brown; *Note:* Very heavy		5, 8, 13	14, 17	19	22	27	30	34
Lovoa trichilioides (L. klaineana) *Common name:* African walnut; *Local names:* Benin, Ghana or Nigerian walnut (UK), dibétou (France, Ivory Coast), apopo (Nigeria), bibolo (Cameroons), eyan (Gabon), nvero or embero (Spanish Guinea), noyer d'Afrique, de Gabon (France, French West Africa); *Source of origin:* West Africa; *Colour:* Lt. gold brown		2, 3, 4, 5, 6, 7	15, 16, 17	20	25	29	32	34
Magnolia grandiflora, M. virginiana *Common name:* Magnolia; *Source of origin:* USA; *Colour:* Yellowish brown		4, 6, 11	18	20	25		33	
Malus sylvestris (M. pumila) *Common name:* Apple; *Sources of origin:* Europe including British Isles				19				
Mangifera salomonensis *Common name:* Pacific walnut (Australia); *Local name:* Island walnut (Australia); *Source of origin:* Solomon Islands; *Colour:* Cream dk. streaky brown	A	4, 6		21			33	

Species	Mode of use	Typical uses	Texture	Density	Strength	Movement	Durability	Permeability
	Uses		*Properties*					
Manilkara zapota *Common name:* Sapodilla; *Source of origin:* Central America; *Note:* Very heavy				19				
Mansonia altissima *Common name:* Mansonia; *Local names:* bété (France, Ivory Coast), aprono (Ghana), ofun (Nigeria); *Source of origin:* West Africa: *Colour:* dk. brown-grey (purple when fresh)	A	3, 4, 5, 6	16, 17, 18	20	24	28	30	34
Measopsis eminii *Common name:* Musizi; *Source of origin:* Tropical Africa				20		29	33	37
Melia composita *Common name:* Lunumidella; *Local names:* Ceylon cedar, Ceylon mahogany (UK); *Source of origin:* Sri Lanka; *Note:* Very light					21			
Mellaleuca leucadendron, M. quinquineruia *Common name:* Tea-tree, *Source of origin:* Australia; *Colour:* Pale pink		1, 5, 7,	16	19			32	
Microberlinia brazzavillensis, M. bisulcata *Common name:* Zebrano; *Local name:* Zingana (France, Gabon); *Source of origin:* West Africa				19				
Millettia laurentii *Common name:* Wenge; *Source of origin:* Zaire				19				
Millettia stuhlmannii *Common name:* Panga panga; *Source of origin:* East Africa				19		29	30	34
Mimusops djave, see ***Baillonella toxisperma***								
Mimusops heckelii, see ***Tieghemella heckelii***								
Mitragyna ciliata *Common name:* Abura (Nigeria); *Local names:* Subaha (Ghana), bahia (Ivory Coast); *Source of origin:* West Africa; *Colour:* Lt. pink-brown	A	4, 5, 6, 12	16, 18	20	25	29		34, 36
Mitragyna stipulosa, M. rubrostipulata *Common name:* Nzingu; *Source of origin:* Uganda				20				
Monopetalanthus heitzii *Common names:* Andoung (France, Gabon) (may include other species); *Source of origin:* West Africa								
Mora excelsa *Common name:* Mora (Morabukea (Guyana) is *M. gonggrijpii); Sources of origin:* Guyana, Trinidad; *Note:* Very heavy				19		27		
Mora gonggrijpii *Common name:* Morabukea; *Source of origin:* Guyana; *Note:* Very heavy				19		27	31	34
Nauclea diderrichii (Sarcocephalus diderrichii) *Common name:* Opepe (Nigeria); *Local names:* Kusia (Ghana), badi (Ivory Coast), bilinga (France, Gabon); *Source of origin:* West Africa; *Colour:* Yellow-gold brown		1, 2, 3, 4, 5, 7, 8, 12, 13	15, 17	19	24	29	30	36
Nauclea orientalis *Common name:* Kanluang; *Sources of origin:* Thailand, Burma								
Nectandra spp. *Common name:* Louro preto (includes *Ocotea* spp.); *Source of origin:* Tropical Africa				19		28	32	34

Species	Uses: Mode of use	Uses: Typical uses	Properties: Texture	Density	Strength	Movement	Durability	Permeability
Nesogordonia papaverifera *Common name:* Danta (Ghana); *Local names:* Otutu (Nigeria), kotibé (France, Ivory Coast); *Source of origin:* West Africa; *Colour:* Pink-brown, dk. red-brown	A, I	1, 3, 4, 5, 6, 8, 12, 13	16, 17	19	24	28	32	36
Newtonia buchananii *Common name:* Muchenche; *Source of origin:* Uganda				20		28	33	34
Nothofagus cunninghamii *Common name:* Tasmanian myrtle; *Local names:* Myrtle beech (Australia), Tasmanian beech; *Source of origin:* Australia; *Colour:* Red-brown	A	5, 6	16	19			32	
Nothofagus dombeyi *Common name:* Coigue; *Local name:* Coihue; *Source of origin:* Chile				20				
Nothofagus menziesii *Common name:* Silver beech; *Local name:* Southland beech (New Zealand); *Source of origin:* New Zealand; *Colour:* Pink-lt. brown	A	4, 5, 6, 11	18	20			33	34
Nothofagus moorei *Common name:* Negro-head beech; *Local name:* Antartic beech (Australia); *Source of origin:* Australia				19				
Nothofagus procera *Common name:* Rauli; *Source of origin:* Chile				20				
Nyssa aquatica *Common name:* Tupelo (includes *N. ogeche, N. sylvatica*); *Local names:* Water tupelo, tupelo gum (USA); *Source of origin:* USA; *Colour:* Brown-grey	A, I	5, 6, 11, 12, 13	17	20	24		33	
Nyssa ogeche *Common names:* Tupelo (includes *N. aquatica, N. sylvatica*); *Source of origin:* USA; *Colour:* Brown-grey	A, I	5, 6, 11, 12, 13	17	20	24		33	
Nyssa sylvatica *Common name:* Tupelo (includes *N. aquatica, N. ogeche*); *Local names:* Black gum, black tupelo (USA); *Source of origin:* USA; *Colour:* Brown-grey	A, I	5, 6, 11, 12, 13	17	20	24		33	
Ochroma pyramidale (O. lagopus, O. bicolor) *Common name:* Balsa; *Sources of origin:* West Indies, Central and South tropical America; *Note:* Very light				21		29	33	35
***Ocotea* spp.** *Common name:* Louro preto (includes *Nectandra* spp.); *Source of origin:* Tropical Africa				19				34
Ocotea barcellensis *Common name:* Louro inamui: *Local name:* Louro inamuhy (Brazil); *Source of origin:* Brazil				19				
Ocotea rodiaei *Common name:* Greenheart; *Source of origin:* Guyana; *Note:* Very heavy				19		28	30	34
Ocotea rubra *Common name:* Red louro; *Local name:* Determa (Guyana), wana (Surinam), louro vermelho (Brazil); *Sources of origin:* Guyana, Surinam, French Guiana, Brazil				20			31	
Ocotea usambarensis *Common name:* East African camphorwood; *Sources: of origin:* Kenya, Tanzania				20		29	30	34

Species	Uses: Mode of use	Uses: Typical uses	Properties: Texture	Density	Strength	Movement	Durability	Permeability
Octomeles sumatrana *Common name:* Binuang (Sarawak, Indonesia, Sabah); *Local names:* Erima, ilimo (New Guinea); *Sources of origin:* Sabah, Sarawak, Indonesia, New Guinea; *Colour:* Grey yellow-lt. brown				21		29	33	36
Olea hochstetteri *Common name:* East African olive; *Local name:* Musheragi (Kenya); *Source of origin:* East Africa; *Note:* Very heavy	A	5	16	19		27	32	36
Olea welwitschii *Common name:* Loliondo (Tanzania); *Local name:* Elgon olive (Kenya); *Source of origin:* East Africa				19		29		
Ostrya carpinifolia *Common name:* European hop-hornbeam; *Sources of origin:* Southern Europe, South-West Asia; *Note:* Very heavy				19				
Ostrya virginiana *Common name:* American hop-hornbeam; *Local names:* Eastern hop-hornbeam (USA), ironwood (USA, Canada), hop-hornbeam (Canada); *Sources of origin:* Canada, USA; *Colour:* Whitish-lt. brown		6		19	24		33	
Oxandra lanceolata *Common name:* Lancewood; *Local name:* Asta (USA); *Source of origin:* West Indies; *Note:* Very heavy				19				
Oxystigma oxyphyllum (pterygopodium oxyphyllum) *Common name:* Tchitola (France, Zaire); *Local names:* Lolagbola (Nigeria), kitola (Zaire), tola, tola manfuta, tola chimfuta (Angola) (not to be confused with agba—*Gossweilerodendron*); *Source of origin:* West Africa				20				
Palaquium spp. *Common name:* Nyatoh; *Local name:* Njatuh (Indonesia), padang (UK); *Sources of origin:* Malaysia, Indonesia		6		19				
Parashorea lucida *Common names:* Meranti gerutu, gerutu gerutu (includes other species); *Source of origin:* West Malaysia				19				
Parashorea malaanonan (P. plicata), P. tomentella *Common name:* White seraya, white lauan (includes lightweight *Pentacme, Shorea* spp.); *Local names:* Urat mata (Sabah), bagtikan (Philippines); *Sources of origin:* Sabah, Philippines; *Colour:* Lt. brown	A, I	4, 5	15	20		29	33	34
Parashorea stellata *Common name:* Thingadu; *Source of origin:* Burma				19			32	
Paratecoma peroba *Common name:* White peroba; *Local names:* Peroba de campos, ipé peroba, peroba amarella, peroba branca (not to be confused with peroba rosa—*Aspidosperma*); *Source of origin:* Brazil				19			30	35
Parinari excelsa *Common name:* Mubura; *Sources of origin:* East and West Africa				19		27	33	36
Parishia insignis *Common name:* Red dhup; *Source of origin:* Andaman Islands				20				
Peltogyne spp. *Common name:* Purpleheart; *Local name:* Amaranth (USA); *Sources of origin:* West Indies, tropical Central and South America				19		29	30	34
Pentace burmanica *Common name:* Thitka; *Local name:* Kashit (Burma); *Source of origin:* Burma				19			31	

Species	Uses: Mode of use	Uses: Typical uses	Properties: Texture	Properties: Density	Properties: Strength	Properties: Movement	Properties: Durability	Properties: Permeability
Pentacme contorta *Common name:* White lauan (includes lightweight *Parashorea, Shorea*); *Local names:* White lauan (Philippines), lamao (Philippines); *Source of origin:* Philippines				20			32	
Pentacme mindanensis *Common name:* White lauan (includes lightweight *Parashorea, Shorea*); *Local name:* Mindanao white lauan (Philippines); *Source of origin:* Philippines				20				
Pericopsis elata *Common name:* Afrormosia; *Local names:* Kokrodua (Ghana, Ivory Coast, France), asamela (Ivory Coast); *Source of origin:* West Africa		2, 3, 4, 5	15	19		29	30	34
Persea lingue *Common name:* Lingue (not to be confused with lingué—*Afzelia*); *Source of origin:* Chile				20				
Phoebe porosa *Common name:* Imbuia; *Local names:* Embuia (Brazil), Brazilian walnut (USA, UK); *Source of origin:* Brazil; *Colour:* Dk. brown		4, 6	15	19		29		
Phyllostylon brasiliensis *Common name:* San Domingo boxwood; *Local name:* Baitoa (Dominican Republic) (boxwoods include *Buxus, Gonioma, Gossypiospermum*); *Sources of origin:* West Indies, Mexico, South America; *Note:* Very heavy				19				
Piptadeniastrum africanum (Piptadenia africana) *Common name:* Dahoma (Ghana); *Local names:* Ekhimi or agboin (Nigeria), dabéma (France, Ivory Coast), mpewere* (Uganda) (* includes muchenche, mkufi—*Newtonia buchananii); Sources of origin:* West and East Africa; *Colour:* Golden brown		1, 5, 7, 8, 12, 13	14, 17	19	24, 25	28	32	35
Piratinera guianensis *Common name:* Snakewood; *Local names:* Letterwood (UK), amourette (France); *Sources of origin:* Tropical central and South America: *Note:* Very heavy				19			30	
***Platanus* spp.—** *Common names:* Plane (English), Platane (French), Plataan (Dutch), Platane (German)								
Platanus hybrida (P. acerifolia) *Common name:* European plane: *Local names:* English, French, etc. plane according to origin, London plane (UK), lacewood (quartered only) (in north of England and Scotland the sycamore—*Acer* is called the plane); *Sources of origin:* Europe, including British Isles				20			33	
Platanus occidentalis *Common name:* American plane; *Local names:* Buttonwood (USA), sycamore* (USA) (*not to be confused with *Acer* spp.); *Source of origin:* USA; *Colour:* Reddish-brown		4, 6, 11, 12	17	20	25		33	
Plathymenia reticulata *Common name:* Vinhatico; *Local name:* Vinhatico castanho (Brazil); *Source of origin:* Brazil				20				
Poga oleosa *Common name:* Poga; *Local names:* Inoi nut (Nigeria), ngale (Cameroons), afo (Spanish Guinea), ovoga (France, Gabon); *Source of origin:* West Africa				21				
Pometia pinata *Common name:* Taun; *Local name:* Ohabu; *Sources of origin:* Papua New Guinea, Malaysia, Indonesia; *Colour:* Pink-red	A	1, 4, 6, 7	18	19	24			

Species	Uses: Mode of use	Uses: Typical uses	Properties: Texture	Properties: Density	Properties: Strength	Properties: Movement	Properties: Durability	Properties: Permeability
Populus spp.— *Common names:* Poplar (English), Peuplier (French), Populier (Dutch), Pappel (German)								
Populus alba *Common name:* White poplar; *Local name:* Abele (UK); *Source of origin:* British Isles; *Colour:* Whitish			18	21				
Populus balsamifera (P. tacamahaca) *Common name:* Canadian poplar (includes *P. grandidentata*); *Local names:* Cottonwood (Canada, USA), balm or black poplar (Canada), balsam poplar (Canada, USA) tacamahac poplar (USA); *Sources of origin:* Canada, USA; *Colour:* White-lt. brown	A, I	6, 11	18	21				
Populus canadensis, var. *serotina* *Common name:* Black poplar (includes *P. nigra, P. robusta*); *Local name:* Black Italian poplar (UK); *Sources of origin:* Europe, including British Isles				21		28	33	36
Populus canescens *Common name:* Grey poplar; *Source of origin:* British Isles			18	20			33	36
Populus deltoides *Common name:* Eastern cottonwood; *Sources of origin:* Eastern Canada and USA; *Colour:* White-lt. brown	A, I	6, 11	18	21	25		33	
Populus grandidentata *Common name:* Canadian poplar (includes *P. balsamifera*); *Local names:* Aspen* (Canada, USA), large tooth aspen (Canada) (* includes *P. tremuloides*); *Sources of origin:* Canada, USA; *Colour:* White-lt. brown	A, I	6, 11	18	21				
Populus nigra *Common name:* Black poplar (includes *P. canadensis,* var. *serotina, P. robusta);* *Local name:* European black poplar (UK); *Sources of Origin:* Europe, including British Isles				21				
Populus robusta *Common name:* Black poplar (includes *P. nigra, P. canadensis,* var. *serotina);* *Local name:* Robusta (UK); *Sources of origin:* Europe, including British Isles				21				
Populus tremula *Common name:* European aspen; *Local names:* Finnish, Swedish, etc. aspen according to origin; *Sources of origin:* Europe, British Isles				21				
Populus tremuloides *Common name:* Canadian aspen; *Local names:* Aspen* (Canada, USA), quaking aspen (Canada, USA), trembling aspen (Canada) (* includes *P. grandidentata*); *Sources of origin:* Canada, USA; *Colour:* White-lt. brown	A, I	11	18	21			33	35
Populus trichocarpa *Common name:* Black cottonwood; *Local names:* (Western balm or balsam cottonwood (Canada, USA), western balsam poplar (Canada);. *Sources of origin:* Western Canada and USA; *Colour:* White-lt. brown	A, I	6, 11	18	21				
Prioria copaifera *Common name:* Cativo; *Local name:* Cautivo (Panama, USA); *Sources of origin:* West Indies, Central America				21				
Prunus spp.— *Common names:* Cherry (English), Merisier (French), Kers (Dutch), Kirsche (German)								
Prunus avium *Common name:* European cherry; *Local name:* Gean or wild cherry (UK); *Source of origin:* Europe, including British Isles; *Colour:* Red-brown		4	16	20		28		

	Uses		Properties					
	Mode of use	Typical uses	Texture	Density	Strength	Movement	Durability	Permeability
Prunus serotina *Common name:* American cherry; *Local names:* Black cherry (Canada, USA), cabinet cherry (USA); *Sources of origin:* Canada, eastern USA; *Colour:* Reddish brown		6	18	20	24		31	
Pseudosindora palustris *Common name:* Swamp sepetir; *Local name:* Sepetir paya, petir (sepetir is *Sindora* spp.); *Source of origin:* Sarawak				19		29		34
Pseudoweinmannia lachnocarpa *Common name:* Mararie; *Source of origin:* Australia; *Colour:* Pink-mauve		12	16	19				
Pterocarpus angolensis *Common name:* Muninga; *Local names:* Mninga (Tanzania), ambila (Mozambique), mukwa (Zambia), kiaat, kajat or kajatenhout (South Africa); *Sources of origin:* Tanzania, Zambia, Angola, Mozambique, Rhodesia, South Africa; *Colour:* Dk. brown		3, 4, 5	15	20		29	30	35
Pterocarpus dalbergioides *Common name:* Andaman padauk; *Local names:* Padauk (UK), Andaman redwood (USA), vermilion wood (USA); *Source of origin:* Andaman Islands; *Colour:* Dk. red		3, 4	15	19		29	30	36
Pterocarpus indicus *Common name:* Amboyna (figured burr wood only); *Local names:* Narra (Philippines, USA), New Guinea rosewood (New Guinea); *Source of origin:* East Indies; *Colour:* Reddish	I	6		19	25		30	
Pterocarpus macrocarpus *Common name:* Burma padauk; *Local names:* Pradoo, mai pradoo (Thailand); *Sources of origin:* Burma, Thailand; *Colour:* Dk. red		3, 4	15	19		29	30	34
Pterocarpus pedatus *Common names:* Maidu, mai dou (Laos); *Local names:* Dang huong (Vietnam), thnong (Khymer Republic), false amboyna (UK); *Sources of origin:* Khymer Republic, Laos, Vietnam; *Note:* Very heavy				19				
Pterocarpus soyauxii, P. osun *Common name:* African padauk; *Local names:* Camwood or barwood (UK), padauk (France), corail (Belgium); *Source of origin:* West Africa; *Colour:* Dk. red		3, 4	15	19		29	30	
Pterocymbium beccarii *Common name:* Amberoi; *Source of origin:* Papua New Guinea; *Colour:* White (radial silver figure)	A	4		21	25		33	
Pterygopodium spp., see ***Oxystigma*** spp.								
Pterygota bequaertii, P. macrocarpa *Common name:* African pterygota; *Local names:* Koto (France, Ivory Coast), ware, awari (Ghana), kefe (Nigeria); *Source of origin:* West Africa				19, 20		28	33	37
Pycnanthus angolensis *Common name:* Ilomba (France, Cameroons, Gabon); *Local name:* Akomu (Nigeria), otie (Ghana) walélé (Ivory Coast), eteng (Cameroons, Gabon); *Source of origin:* West and East Africa				20			33	37
Pyrus communis *Common name:* Pear; *Sources of origin:* Europe, including British Isles				19				
Pyrus malus, see ***Malus sylvestris***								
Quercus spp.— *Common names:* Oak (English), Chêne (French), Eik (Dutch), Eiche (German)								

Species	Uses: Mode of use	Uses: Typical uses	Properties: Texture	Properties: Density	Properties: Strength	Properties: Movement	Properties: Durability	Properties: Permeability
Quercus castaneaefolia *Common name:* Persian oak; *Source of origin:* Iran				19				
Quercus cerris *Common name:* Turkey oak; *Sources of origin:* Europe, including British Isles				19		27	32	34
Quercus ilex *Common name:* Holm oak; *Local name:* Evergreen oak (UK); *Sources of origin:* Europe, including British Isles				19			30	
Quercus mongolica var. ***grosseserrata (Q. crispula, Q. grosseserrata)*** *Common name:* Japanese oak; *Local names:* Ohnara (*Q. mongolica*), konara (*Q. glandulifera*), kashiwa (*Q. dentata*); *Source of origin:* Japan		3, 4, 5, 6, 8		19		28	33	
Quercus petraea (Q. sessiliflora) *Common name:* European oak (includes *Q. robur*); *Local names:* English, French, Polish, etc. according to origin, sessile oak, durmast oak (UK); *Sources of origin:* Europe, including British Isles; *Colour:* Lt. brown	A	3, 4, 5, 7, 8, 10, 12	14	19		28	31	34
Quercus robur (Q. pedunculata) *Common name:* European oak (includes *Q. petraea*); *Local names:* English, French, Polish oak according to origin, pedunculate oak (UK); *Sources of origin:* Europe, including British Isles; *Colour:* Lt. brown	A	3, 4, 5, 7, 8, 10, 12	14	19		28	31	34, 36
Quercus spp. *Common names:* American red oak, includes *Q. rubra* (*Q. borealis*), northern red oak (Canada, USA), *Q. falcata* var. *falcata*, southern red or Spanish oak (USA), *Q. falcata* var. *pagodaefolia*, swamp red oak, cherrybark oak (USA), *Q. shumardii*, shumard red oak (USA); *Sources of origin:* Eastern Canada and USA; *Colour:* Lt. reddish brown		4, 5, 6, 11, 12	18	19	24	28	33	35
Quercus spp. *Common names:* American white oak, includes *Q. alba*, white oak (USA), *Q. prinus* (*Q. montana*), chestnut oak (USA), *Q. lyrata*, overcup oak (USA), *Q. michauxii*, swamp chestnut oak (USA); *Sources of origin:* Eastern Canada and USA; *Colour:* Lt. brown		3, 4, 5, 6, 7, 9, 10, 11, 12, 13	18	19	23	28	31	34
Rapanea rhododendroides *Common name:* Rapanea; *Local names:* Mlimangombe (Tanzania), mugaita (Kenya); *Source of origin:* East Africa; *Note:* Very heavy				19				
Ricinodendron heudelotii (R. africanum) *Common name:* Erimado; *Local names:* Wama (Ghana), sanga sanga (Zaire), erimado (Nigeria), essessang (France, Cameroons, Gabon), musodo (Uganda); *Source of origin:* Tropical Africa; *Note:* Very light				21				
Ricinodendron rautanenii *Common name:* Mugongo; *Local names:* Mgongo (Rhodesia), mungongo* (Botswana) (also includes in Zaire *Antrocaryon* spp., onzabili); *Sources of origin:* South tropical Africa, Rhodesia, Botswana; *Note:* Very light				21				
Robinia pseudoacacia *Common name:* Robinia; *Local names:* Black locust (USA), false acacia (UK); *Sources of origin:* Europe, including British Isles, USA: *Colour;* Yellowish-golden brown		9, 10, 11, 13		19	22		31	
Salix spp.— *Common names:* Willow (English), Saule (French), Wilg (Dutch), Weide (German)								
Salix alba, S. fragilis and hybrids *Common name:* White willow; *Local name:* Common willow (UK); *Source of origin:* British Isles				21			33	35

Species	Uses: Mode of use	Uses: Typical uses	Properties: Texture	Properties: Density	Properties: Strength	Properties: Movement	Properties: Durability	Properties: Permeability
Salix alba, **var. *calva*** *Common name:* Cricket-bat willow; *Local name:* Close-bark willow (UK)				21				
Salix nigra *Common name:* Black willow; *Source of origin:* USA; *Colour:* lt. brown		6, 11	18	21	26		33	
Sandoricum indicum *Common name:* Katon (Thailand); *Local name:* Thitto (Burma); *Sources of origin:* Burma, Thailand, East Indies				20			32	
Santalum album *Common name:* Sandalwood; *Source of origin:* India; *Note:* Very heavy				19				
Santalum spicatum *Common name:* Sandalwood (Australia); *Source of origin:* Australia								
***Sarcocephalus* spp., see *Nauclea* spp.**								
Schizomeria ovata *Common name:* White birch (includes *S. whitei,* North Queensland); *Local names:* Crab apple (New South Wales), New South Wales white ash; *Source of origin:* Australia; *Colour:* White-lt. brown	A, I			20			33	
Scottellia coriacea *Common name:* Odoko; *Source of origin:* West Africa				20		28	33	37
***Shorea* spp. (from West Malaysia)** *Common names:* Light red meranti, red meranti—principally *S. acuminata, S. leprosula, S. macroptera, S. ovalis, S. parvifolia; Source of origin:* West Malaysia; *Colour:* Pale red or pink				20			31	34
***Shorea* spp. (from Sarawak, Brunei)** *Common name:* Light red meranti, red meranti (Sarawak), perawan (Sarawak), meranti bunga (Sarawak), alan bunga (heavier, *S. albida* only)—principally *S. albida, S. parvifolia, S. quadrinervis*; *Source of origin:* Sarawak, Brunei; *Colour:* Pale red or pink				20			31	34
***Shorea* spp. (from Sabah)** *Common names:* Light red seraya, red seraya (Sabah), seraya merah (Sabah)—principally *S. leprosula, S. leptoclados, S. parvifolia, S. smithiana*; *Source of origin:* Sabah; *Colour:* Pale red or pink	A, I	2, 3, 4, 5	16	20		29	31	34
***Shorea* spp. (from Indonesia)** *Common names:* Light red meranti, red meranti (Indonesia), lanan (Kalimantan)—principally *S. leprosula, S. ovalis, S. parvifolia*; *Sources of origin:* Sumatra, Kalimantan; *Colour:* Pale red or pink	A, I	2, 3, 4, 5	16	20		29	31	34
***Shorea* spp. (from Philippines)** *Common names:* Red lauan, dark red lauan, Philippine mahogany—principally *S. polysperma* (red lauan), *S. squamata (S. palosapis)* (tangile, bataan, mayapis) *S. agsaboensis* (tiaong); *Source of origin:* Philippines; *Colour:* Dk. red				20				
Shorea almon (S. eximia), S. squamata (S. palosapis) *Common name:* White lauan (also includes *Parashorea, Pentacme), Local names:* light red lauan (UK), Philippine white mahogany, *S. almon,* almon (Philippines), *S. squamata,* mayapis (Philippines); *Source of origin:* Philippines; *Colour:* Pale pink or white				20				
***Shorea* spp. (from West Malaysia, Sarawak, Brunei)** *Common names:* Dark red meranti; nemesu, *S. pauciflora* only (West Malaysia)—principally *S. pauciflora, S. acuminata, S. curtisii*; *Sources of origin:* West Malaysia, Sarawak, Brunei; *Colour:* Dk. red				19				

	Uses		Properties					
	Mode of use	*Typical uses*	*Texture*	*Density*	*Strength*	*Movement*	*Durability*	*Permeability*
***Shorea* spp. (from West Malaysia, Sarawak, Brunei)** *Common names:* Yellow meranti, lun, lun kuning (Sarawak)—principally S. *faguetiana, S. multiflora, S. resina-nigra* (Malaysia), *S. hopeifolia* (Sarawak, Brunei); *Sources of origin:* West Malaysia, Sarawak, Brunei; *Colour:* Yellow				19		28	32	34
***Shorea* spp. (from Sabah)** *Common names:* Dark red seraya, oba suluk (Sabah)—principally *S. pauciflora*; *Source of origin:* Sabah; *Colour:* Dk. red				19			32	34
***Shorea* spp. (from Sabah)** *Common names:* Yellow seraya, seraya kuning (Sabah)—principally *S. acuminatissima, S. gibbosa, S. faguetiana*; *Source of origin:* Sabah; *Colour:* Yellow				19			32	34
***Shorea* spp. (from West Malaysia, Sarawak, Brunei, Sabah)** *Common names:* White meranti, lun, lun puteh (Sarawak), melapi (Sabah); *Sources of origin:* West Malaysia, Sarawak, Brunei, Sabah; *Colour:* White, or pale coloured				19				36
Shorea albida *Common name:* Alan; *Local names:* Meranti bunga, alan bunga, meraka alan (Sarawak), red selangan (Sarawak), selangan merah (Sarawak), seringawan (Brunei); *Sources of origin:* Sarawak, Brunei; *Colour:* Red				19				
Shorea guiso *Common names:* Red selangan batu (Sabah), red balau (West Malaysia, includes *S. kunstleri*); *Local name:* Selangan batu merah (Sabah); *Sources of origin:* West Malaysia, Sabah; *Colour:* Red				19		28		34
Shorea robusta *Common name:* Sal; *Source of origin:* India; *Colour:* Yellow or brown				19				34
***Shorea* spp.** *Common name:* Selangan batu—principally *S. laevis, S. seminis, S. superba*; *Sources of origin:* Sarawak, Brunei, Sabah; *Colour:* Yellow or brown; *Note:* Very heavy				19		29		34
***Shorea* spp.** *Common name:* Chan; *Source of origin:* Thailand; *Colour:* Yellow or brown				19				
***Shorea* spp.** *Common name:* Balau—principally *S. glauca, S. maxwelliana; Source of origin:* West Malaysia; *Colour:* Yellow or brown; *Note:* Very heavy				19				34
Simaruba amara *Common name:* Simaruba; *Local names:* Marupa (Brazil), simarupa (Guianas); *Source of origin:* Tropical America				21			32	
Sindora coriacea *Common name:* Sepetir; *Local names:* Makata (Thailand), supa (Philippines), (petir, see *Pseudosindora); Sources of origin:* Vietnam, Khymer Republic, Thailand, Malaysia, Philippines				19			31	36
Sloanea australia *Common name:* Blush alder; *Local name:* Maiden's blush, blush carabeen; *Source of origin:* Australia; *Colour:* Pink-reddish	A, I	1, 5, 6,		19			33	
Sloanea langii *Common name:* White carabeen; *Source of origin:* Australia								
Sloanea macbrydei *Common name:* Grey carabeen; *Source of origin:* Australia								
Sloanea woollsii *Common name:* Yellow carabeen; *Source of origin:* Australia; *Colour:* Pale brown	A	4, 5, 11	15	20				

Species	Uses: Mode of use	Uses: Typical uses	Properties: Texture	Density	Strength	Movement	Durability	Permeability
Sorbus aria *Common name:* Whitebeam; *Sources of origin:* Europe, including British Isles				19				
Staudtia stipitata (C. gabonesis) *Common name:* Niovè; *Source of origin:* West Africa				19			31	
Sterculia oblonga *Common name:* Yellow sterculia; *Local names:* White sterculia (UK), okoko (Nigeria), eyong (Cameroons); *Source of origin:* West Africa				19		28	33	34
Sterculia pruriens *Common name:* Maho; *Source of origin:* Guyana, Brazil				20		27	33	37
Sterculia rhinopetala *Common name:* Brown sterculia; *Local names:* Red sterculia (UK), wawabima (Ghana), aye (Nigeria), lotofa (Ivory Coast); *Source of origin:* West Africa				19		27	32	34
Swartzia leiocalycina *Common name:* Wamara; *Source of origin:* Guyana; *Note:* Very heavy				19		28	31	34
Swietenia candollei *Common name:* American mahogany (includes other species); *Local name:* Venezuelan mahogany; *Source of origin:* Venezuela; *Colour:* Red or gold-brown		5, 6, 7,		19				
Swietenia macrophylla *Common name:* American mahogany (includes other species); *Local names:* Central American mahogany (UK), British Honduras, Costa Rica, Guatemalan, Honduras, Mexican, etc. mahogany according to origin; *Source of origin:* Central America; *Colour:* Red or gold-brown		3, 4, 5, 6, 7	15	20		29	31	34
Swietenia macrophylla(?), S. tessmannii(?) *Common name:* American mahogany (includes other species); *Local names:* Peruvian mahogany, aguano, caoba (usually shipped from Brazil); *Source of origin:* Peru; *Colour:* Red or gold brown		5, 6, 7		20				
Swietenia macrophylla(?), S. krukovii(?) *Common name:* American mahogany (includes other species); *Local names:* Brazilian mahogany, aguano, araputanga, mogno (Peruvian mahogany is also shipped from Brazil); *Source of origin:* Brazil; *Colour:* Red or gold brown		5, 6, 7,		20				
Swietenia mahagoni *Common name:* American mahogany (includes other species); *Local names:* Cuban, Jamaican, Porto Rico, San Domingan, West Indian, etc. mahogany, according to origin; *Source of origin:* West Indies; *Colour:* Red or gold brown		5, 6, 7		19			31	34
Syncarpia glomulifera (S. laurifolia) *Common name:* Turpentine; *Local name:* Luster (when dressed); *Source of origin:* Australia; *Colour:* Red-brown; *Note:* Very heavy		5, 8,		19	23		30	34
Syncarpia hillii *Common name:* Satinay; *Local names:* Red satinay, Fraser Island turpentine; *Source of origin:* Australia (Fraser Island); *Colour:* Pink	A	1, 2, 3, 6, 8, 9	15	19	24		31	
Tabebuia donnell-smithii *Common name:* Prima vera; *Sources of origin:* Mexico, Central America				21		29		
Tarrietia spp., see *Heritiera* for Asian spp. only								
Tarrietia densiflora *Common name:* Niangon (includes *T. utilis*); *Local name:* Ogoué (Gabon); *Source of origin:* West Africa; *Colour:* Lt. red-brown				20				

Species	Uses: Mode of use	Uses: Typical uses	Properties: Texture	Density	Strength	Movement	Durability	Permeability
Tarrietia utilis *Common name:* Niangon (includes *T. densiflora*); *Local names:* Nyankom (Ghana), wishmore (Liberia); *Source of origin:* West Africa; *Colour:* Lt. red-brown		1, 3, 4, 5, 6, 7, 12	15, 17	20	25	28	32	
Tectona grandis *Common name:* Teak; *Sources of origin:* India, Burma, Thailand, Jara; *Colour:* Brown	A	2, 3, 4, 5, 6, 7, 8	14	19		29	30	34
Terminalia alata, T. Coviacea, T. crenulata (formerly T. tomentosa) *Common name:* Indian laurel; *Local names:* Taukkyan (Burma), asna, mutti or sain (India); *Sources of origin:* Burma, India; *Colour:* Dk. brown		4, 6	15	19		28	32	34
Terminalia amazonia *Common name:* Nargusta; *Local names:* White olivier, steud; *Sources of origin:* British Honduras, Trinidad				19		28	30	34
Terminalia arjuna *Common name:* Kumbuk; *Local name:* Arjun (India); *Sources of origin:* India, Sri Lanka				19				
Terminalia bialata *Common names:* White chuglam, Indian silver-grey wood; *Source of origin:* Andaman Islands				19			32	
Terminalia ivorensis *Common name:* Idigbo (Nigeria); *Local names:* Emeri (Ghana), framiré (France, Ivory Coast), black afara (Nigeria); *Source of origin:* West Africa; *Colour:* Yellow		1, 2, 3, 4, 5, 6, 7, 12	15, 17, 18	20	25	29	31	34
Terminalia procera *Common name:* White bombway; *Local name:* Badam (India); *Source of origin:* Andaman Islands				19		29	33	36
Terminalia sericocarpa *Common name:* Damson (Australia); *Local name:* Sovereign wood; *Source of origin:* Australia; *Colour:* Pale gold	A, I	1, 4, 6	14	20			33	
Terminalia superba *Common names:* Afara (Nigeria), Limba (France, Zaire); *Local names:* Fraké (Ivory Coast), ofram (Ghana), white afara (Nigeria), akom (Cameroons, Gabon), noyer du Mayombe, limbo (France); *Source of origin:* West Africa; *Colour:* Yellowish		4	15	20		29	33	36
Tetraberlinia tubmaniana *Common name:* Tetraberlinia; *Source of origin:* Liberia				20		29		
Tetrameles nudiflora *Common name:* Kapong; *Local name:* Sompong (Thailand); *Source of origin:* Thailand; *Note:* Very light				21				
Tetramerista glabra *Common name:* Punah; *Source of origin:* Malaysia				19			33	35
Tieghemella africana *Common name:* Makoré (includes *T. heckelii); Local name:* Douka (Cameroons, Gabon); *Source of origin:* West Africa; *Colour:* Red-brown	A, I	3, 4, 5, 6, 7, 12	16, 18	20	25	27	30	
Tieghemella heckelii *Common name:* Makoré (includes *T. africana*); *Local name:* Baku (Ghana); *Source of origin:* West Africa; *Colour:* Lt. red	A, I	2, 3, 4, 5	16	20		29	30	34

Species	Uses: Mode of use	Uses: Typical uses	Properties: Texture	Properties: Density	Properties: Strength	Properties: Movement	Properties: Durability	Properties: Permeability
Tilia spp.— *Common names:* Lime (English), Tilleul (French), Linde (Dutch), Linde (German								
Tilia americana (T. glabra) *Common name:* Basswood; *Local name:* American lime (UK); *Sources of origin:* Canada, eastern USA; *Colour:* Lt. reddish-brown	I	4, 6, 11	18	21	25		33	37
Tilia japonica *Common name:* Japanese lime; *Local names:* Shina (Japan), shinanoki (Japan), Japanese basswood (UK); *Source of origin:* Japan				21				
Tilia vulgaris (T. europaea) *Common name:* European lime; *Local name:* English lime, etc. according to origin; *Sources of origin:* Europe, including British Isles: *Colour:* Whitish			16	20		28	33	37
Toona australis (Cedrela toona) *Common name:* Australian cedar; *Local name:* Red cedar (Australia); *Source of origin:* Australia; *Colour:* Red-brown	A	6		21			30	
Toona ciliata (Cedrela toona) *Common name:* Burmese cedar; *Local names:* Toon (India, Pakistan), yomhom (Thailand); *Sources of origin:* Burma, India, Pakistan, Thailand				21			32	36
Triplochiton scleroxylon *Common names:* Obeche (Nigeria), wawa (Ghana); *Local names:* Arere (Nigeria), ayous (Cameroons, Zaire), samba (Ivory Coast); *Source of origin:* West Africa; *Colour:* Cream-yellowish	A, I	4, 6, 7, 12	15, 17, 18	21	26	29	33	35
Tristania conferta *Common name:* Brush box; *Source of origin:* Australia; *Colour:* Pink-brown; *Note:* Very heavy	A	5, 8	16	19			32	34
Turraeanthus africanus *Common name:* Avodiré; *Sources of origin:* Ghana, Ivory Coast, Angola				20		29	32	34
Ulmus spp.— *Common names:* Elm (English), Orme (French), Iep (Dutch), Ulme (German)								
Ulmus americana *Common name:* White elm; *Local names:* American elm (Canada, UK), soft elm (USA); *Sources of origin:* Eastern Canada and USA; *Colour:* Lt. brown		6, 11, 12		20	25		33	36
Ulmus carpinifolia (U. nitens) *Common name:* Smooth-leaved elm; *Local names:* French elm, Flemish elm, etc. according to origin; *Sources of origin:* Europe, including British Isles				20				
Ulmus glabra (U. montana) *Common name:* Wych elm; *Local names:* Mountain elm, Scotch elm, white elm (UK); *Source of origin:* British Isles; *Colour:* Lt. brownish		2, 6, 7, 8		19			33	35
Ulmus hollandica, var. *hollandica* *Common name:* Dutch elm; *Local name:* Cork bark elm (UK); *Source of origin:* British Isles; *Colour:* Lt. brownish		2, 6, 8		20			33	36
Ulmus laciniata, U. davidiana, var. *japonica* *Common name:* Japanese elm; *Local name:* Nire (Japan); *Source of origin:* Japan				20				
Ulmus procera (U. campestris) *Common name:* English elm; *Local names:* Red elm, nave elm (UK); *Source of origin:* British Isles; *Colour:* Lt. brownish		2, 6, 8	14	20		28	33	36

Species	Uses: Mode of use	Uses: Typical uses	Properties: Texture	Density	Strength	Movement	Durability	Permeability
Ulmus thomasii (U. racemosa) *Common name:* Rock elm; *Local names:* Canadian rock elm (Canada, UK), cork elm (Canada, USA), cork bark elm (USA), hickory elm (USA); *Sources of origin:* Eastern Canada and USA; *Colour:* Lt. brown		6, 7, 11, 12		19	23		33	36
Virola bicuhyba *Common name:* Heavy virola; *Local name:* Bicuiba (Brazil); *Source of origin:* Brazil				19				
Virola koschnyi *Common name:* Light virola (includes other species); *Local names:* Banak, sangre palo, palo de sangre (British Honduras), tapsava (USA); *Sources of origin:* Tropical Central and South America				20			33	37
Virola sebifera, V. surinamensis *Common name:* Light virola (includes *V. koschnyi); Local names:* Dalli (Guyana), baboen (Surinam), ucuumba, virola (Brazil); *Sources of origin:* Tropical Central and South America				20				
Vitex cofassus *Common name:* New Guinea teak; *Source of origin:* Papua New Guinea; *Colour:* Lt. brown	A	1, 3, 4, 6, 7, 10		19	24	29	31	
Vochysia hondurensis, V. guianensis, V. tetraphylla *Common name:* Quaruba (Brazil); *Local names:* Yemeri (British Honduras), cedro-rana (Brazil), iteballi (Guyana), kwarie (Surinam); *Sources of origin:* Tropical Central and South America				20			33	37
***Xanthostemon* spp.** *Common name:* Penda (includes *Tristania pachysperma*); *Local names:* Brown or Johnstone river penda (*X. chrysanthus*), red or Atherton penda (*X. pubescens),* southern penda *(X. oppositifolius),* yellow penda or Johnstone river yellowwood *(T. pachysperma) (T. pachysperma* not so durable or strong); *Source of origin:* Australia; *Note:* Brown, red and southern penda very heavy	A	1		19				
Xylia dolabriformis *Common name:* Pyinkado (includes *X. xylocarpa); Local name:* Cam xe (Vietnam); *Source of origin:* India; *Note:* Very heavy				19		28	30	34
Xylia xylocarpa *Common name:* Pyinkado (includes *X. dolabriformis*); *Local name:* Viul (India); *Sources of origin:* India, Burma; *Note:* Very heavy				19		28	30	

Index to Common Names

Note:– Entries in this index are arranged in alphabetical order of the noun in the common names of woods. Thus brown beech will be found as "beech, brown" and Philippine mahogany as "mahogany, Philippine". Once the botanical name has been determined it can be used to find the entry in the table in Appendix 3 which lists alternative common names, sources, typical uses and basic properties.

Entries marked § are listed in the first (softwood) part, and all the other entries in the second (hardwood) part of the table in Appendix 3. Entries marked ‖ show a variation of name indicating the geographical source of the wood. If a wood is described in the text it is listed in the General Index under the preferred common name.

This index, in its choice of preferred common names, conforms with British Standards 589 and 881, but contains numerous additions to cover woods commonly used in North America, South Africa and Australasia.

anokye, *Guibourtia ehie*
antiaris, *Antiaris toxicaria*
anyaran, *Distemonanthus benthamianus*
apa, *Afzelia* spp.
apitong, *Dipterocarpus* spp.
apopo, *Lovoa trichilioides*
apple, *Malus sylvestris*
–, crab, *Schizomeria ovata*
–, star, *Chrysophyllum cainito*
aprono, *Mansonia altissima*
araputanga, *Swietenia macrophylla(?)*
arborvitae, eastern, *Thuja occidentalis* §
–, giant, *Thuja plicata* §
arere, *Triplochiton scleroxylon*
arjun, *Terminalia arjuna*
asamela, *Pericopsis elata*
ash, alpine, *Eucalyptus delegatensis*
–, American, *Fraxinus americana*
–, Australian white, *Eucalyptus fraxinoides*
–, black, *Fraxinus nigra*
–, – mountain, *Eucalyptus fastigata*
–, brown, *Fraxinus nigra*
–, Canadian, *Fraxinus americana*
–, canary, *Beilschmiedia bancroftii*
–, crow's, *Flindersia australis*
–, English, *Fraxinus excelsior*
–, European, *Fraxinus excelsior*
–, French, *Fraxinus excelsior*
–, green, *Fraxinus pennsylvanica*
–, hickory, *Flindersia ifflaiana*
–, Japanese, *Fraxinus mandshurica*
–, mountain, *Eucalyptus regnans*
–, Polish, *Fraxinus excelsior*
–, red, *Fraxinus pennsylvanica*
–, rose, *Flindersia laevicarpa*
–, silver, *Flindersia* spp.
–, silver-top, *Eucalyptus sieberi*
–, Slavonian, *Fraxinus excelsior*
–, Victorian, *Eucalyptus regnans*
–, white, *Eucalyptus fraxinoides*
–, white, *Fraxinus americana*
–, white, *Schizomeria ovata*
asna, *Terminalia alata*
aspen, *Populus grandidentata*
–, *Populus tremuloides*
–, Canadian, *Populus tremuloides*
–, European, *Populus tremula*
–, Finnish, *Populus tremula*
–, large tooth, *Populus grandidentata*
–, quaking, *Populus tremuloides*
–, Swedish, *Populus tremula*
–, trembling, *Populus tremuloides*
assacu, *Hura crepitans*
assie, *Entandrophragma utile*
asta, *Oxandra lanceolata*
aune, *Alnus* spp.
avodiré, *Turraeanthus africanus*
awari, *Pterygota bequaertii*
awun, *Alstonia congensis*
ayan, *Distemonanthus benthamianus*
aye, *Sterculia rhinopetala*
ayinre, *Albizia* spp.
ayous, *Triplochiton scleroxylon*
azobé, *Lophira alata*

baboen, *Virola surinamensis*
bacu, *Cariniana* spp.
badam, *Terminalia procera*
badi, *Nauclea diderrichii*
bagac, *Dipterocarpus* spp.
bagtikan, *Parashorea malaanonan*
bahia, *Mitragyna ciliata*
– wood, *Caesalpinia echinata*
baitoa, *Phyllostylon brasiliensis*
baku, *Tieghemella heckelii*
balau, *Shorea* spp. (heavy, yellow-brown)
–, red, *Shorea* spp. (heavy, red)
balsa, *Ochroma pyramidale*
balsam, *Abies balsamea* §
banak, *Virola koschnyi*
banga wanga, *Amblygonocarpus andogensis*
bangalay, *Eucalyptus botryoides*
banglang, *Lagerstroemia speciosa*
baromalli, *Catostemma commune*
barrel, brown, *Eucalyptus fastigata*
barwood, *Pterocarpus soyauxii*
basralocus, *Dicorynia guianensis*
basswood, *Tilia americana*
–, *Endospermum* spp.
bataan, *Shorea polysperma* (medium weight, dark red)
baywood, *Swietenia macrophylla*
bean, black, *Castanospermum australe*
–, red, *Geissois benthami*
–, walnut, *Endiandra palmerstonii*
–, yellow, *Albizia xanthoxylon*
beech, American, *Fagus grandifolia*
–, Antarctic, *Nothofagus moorei*
–, brown, *Cryptocarya glaucescens*
–, brown, *Litsea reticulata*
–, Carpathian, *Fagus sylvatica*
–, Danish, *Fagus sylvatica*
–, English, *Fagus sylvatica*
–, European, *Fagus sylvatica*
–, French, *Fagus sylvatica*
–, Japanese, *Fagus crenata*
–, myrtle, *Nothofagus cunninghamii*
–, negro-head, *Nothofagus moorei*

beech, silver, *Nothofagus menziesii*
–, Slavonian, *Fagus sylvatica*
–, Southland, *Nothofagus menziesii*
–, Tasmanian, *Nothofagus cunninghamii*
–, Turkish, *Fagus orientalis*
belian, *Eusideroxylon zwageri*
bendang, *Agathis dammara* §
benge, *Guibourtia arnoldiana*
Beninwood, *Khaya grandifoliola*
berk, *Betula* spp.
berlinia, *Berlinia* spp.
bété, *Mansonia altissima*
beuk, *Fagus* spp.
bibolo, *Lovoa trichilioides*
bicuiba, *Virola bicuhyba*
bilinga, *Nauclea diderrichii*
billian, *Eusideroxylon zwageri*
bilsted, *Liquidambar styraciflua*
bindang, *Agathis dammara* §
bintangor, *Calophyllum* spp.
binuang, *Octomeles sumatrana*
birch, American, *Betula alleghaniensis*
–, American, *Betula papyrifera*
–, Canadian yellow, *Betula alleghaniensis*
–, English, *Betula* spp.
–, European, *Betula* spp.
–, Finnish, *Betula* spp.
–, hard, *Betula alleghaniensis*
–, Japanese, *Betula maximowieziana*
–, paper, *Betula papyrifera*
–, Quebec, *Betula alleghaniensis*
–, silver, *Betula* spp.
–, Swedish, *Betula* spp.
–, western, *Betula papyrifera* var. *occidentalis*
–, western paper, *Betula papyrifera* var. *occidentalis*
–, western white, *Betula papyrifera* var. *occidentalis*
–, white, *Betula papyrifera*
–, white, *Betula* spp.
–, white, *Schizomeria ovata*
–, yellow, *Betula alleghaniensis*
Birke, *Betula* spp.
bisselon, *Khaya senegalensis*
blackbutt, *Eucalyptus pilularis*
–, pink, *Eucalyptus engenioides,* etc.
–, Western Australian, *Eucalyptus patens*
blackwood, African, *Dalbergia melanoxylon*
–, Australian, *Acacia melanoxylon*
–, Bombay, *Dalbergia latifolia*
bloodwood, red, *Eucalyptus gummifera,* etc.
bois de rose, *Dalbergia frutescens* var. *tomentosa*
bois satiné, *Brosimum paraense*
bombax, West African, *Bombax buonopozense*
bombway, white, *Terminalia procera*
bongossi, *Lophira alata*
bonsamdua, *Distemonanthus benthamianus*
boojong, *Heritiera trifoliolata*
bossé, *Guarea* spp.
bouleau, *Betula* spp.
box, *Buxus sempervirens*
–, brush, *Tristania conferta*
–, Cape, *Buxus macowani*
–, coast grey, *Eucalyptus bosistoana*
–, grey, gum top, *Eucalyptus microcarpa*
–, red, *Eucalyptus polyanthemos*
–, yellow, *Eucalyptus melliodora*
boxwood, Abassian, *Buxus sempervirens*
–, Cape, *Gonioma kamassi*
–, Colombian, *Gossypiospermum praecox*
–, East London, *Buxus macowani*
–, European, *Buxus sempervirens*
–, Iranian, *Buxus sempervirens*
–, kamassi, *Gonioma kamassi*
–, Knysna, *Gonioma kamassi*
–, Maracaibo, *Gossypiospermum praecox*
–, Persian, *Buxus sempervirens*
–, San Domingo, *Phyllostylon brasiliensis*
–, Turkey, *Buxus sempervirens*
–, Venezuelan, *Gossypiospermum praecox*
–, West Indian, *Gossypiospermum praecox*
brachystegia, *Brachystegia* spp.
bubinga, *Guibourtia demeusei*
Buche, *Fagus* spp.
Buchsbaum, *Buxus* spp.
buis, *Buxus* spp.
buna, *Fagus crenata*
butt, woolly, *Eucalyptus delegatensis*
–, woolly, silver or white, *Eucalyptus fastigata*
buttonwood, *Platanus occidentalis*

cabbage bark, *Andira inermis*
–, red, *Andira inermis*
cadaga, *Eucalyptus torelliana*
calophyllum, beech, *Calophyllum inophyllum*
–, bush, *Calophyllum kajewski*
camphorwood, East African, *Ocotea usambarensis*
camwood, *Pterocarpus soyauxii*
canarium, African, *Canarium schweinfurthii*
–, Indian, *Canarium euphyllum*
–, Malaysian, *Canarium* spp.
canarywood, *Liriodendron tulipifera*
caoba, *Swietenia macrophylla(?)*
carabeen, blush, *Sloanea australia*
–, grey, *Sloanea macbrydei*
–, red, *Geissois benthami*
–, white, *Sloanea langii*
–, yellow, *Sloanea woollsii*

cativo, *Prioria copaifera*
cautivo, *Prioria copaifera*
cedar, *Cedrela fissilis*
–, *Entandrophragma cylindricum*
–, *Entandrophragma utile*
–, *Cedrus atlantica* §
–, acacia, *Albizia toona*
–, African pencil, *Juniperus procera* §
–, Alaska yellow-, *Chamaecyparis nootkatensis* §
–, Atlantic, *Cedrus atlantica* §
–, Atlantic white, *Chamaecyparis thyoides* §
–, Atlas, *Cedrus atlantica* §
–, Australian, *Toona australis*
–, Brazilian, *Cedrela fissilis*
–, British Columbia red, *Thuja plicata* §
–, British Guiana, *Cedrela fissilis*
–, British Honduras, *Cedrela odorata*
–, Burma, *Toona ciliata*
–, California incense, *Calocedrus decurrens* §
–, Central American, *Cedrela odorata*
–, Ceylon, *Melia composita*
–, cigar-box, *Cedrela fissilis* and *Cedrela odorata*
–, East African pencil, *Juniperus procera* §
–, eastern red, *Juniperus viginiana* §
–, eastern white, *Thuja occidentalis* §
–, Honduras, *Cedrela odorata*
–, incense, *Calocedrus decurrens* §
–, incense, California, *Calocedrus decurrens* §
–, Japanese, *Cryptomeria japonica* §
–, Mexican, *Cedrela odorata*
–, Nicaraguan, *Cedrela odorata*
–, Nigerian, *Gossweilerodendron balsamiferum*
–, Nigerian, *Guarea cedrata*
–, northern white, *Thuja occidentalis* §
–, Pacific Coast yellow, *Chamaecyparis nootkatensis* §
–, pencil, *Ackama australiensis, paniculata*
–, pencil, *Dysoxylum muelleri*
–, Peruvian, *Cedrela fissilis*
–, Port Orford, *Chamaecyparis lawsoniana* §
–, red, *Toona australis*
–, red, *Thuja plicata* §
–, Siberian, *Pinus sibirica* §
–, South American, *Cedrela fissilis*
–, southern white, *Chamaecyparis thyoides* §
–, Tabasco, *Cedrela odorata*
–, Trinidad, *Cedrela odorata*
–, Virginian pencil, *Juniperus virginiana* §
–, western red, *Thuja plicata* §
–, West Indian, *Cedrela odorata*
–, white, *Chamaecyparis thyoides* §
–, white, *Thuja occidentalis* §
–, yellow, *Chamaecyparis nootkatensis* §
cedar of Lebanon, *Cedrus libani* §
ceder, *Cedrus* spp. §
cèdre, *Cedrus* spp. §
cedro, *Cedrela fissilis*
cedro-rana, *Vochysia* spp.
ceiba, *Ceiba pentandra*
celtis, African, *Celtis* spp.
chamfuta, *Afzelia quanzensis*
chan, *Shorea* spp. (heavy, yellow-brown)
charme, *Carpinus betulus*
châtaignier, *Castanea* spp.
cheesewood, white, *Alstonia scholaris*
chenchen, *Antiaris africana*
chêne, *Quercus* spp.
cherry, *Prunus avium*
–, American, *Prunus serotina*
–, black, *Prunus serotina*
–, cabinet, *Prunus serotina*
–, European, *Prunus avium*
–, wild, *Prunus avium*
chestnut, American, *Castanea sativa*
–, European, *Castanea sativa*
–, Spanish, *Castanea sativa*
–, sweet, *Castanea sativa*
chestnut oak, *Quercus prinus*
– –, swamp, *Quercus falcata*
chickrassy, *Chukrasia tabularis*
chittagong wood, *Chukrasia tabularis*
chuglam, white, *Terminalia bialata*
chumprag, *Heritiera cochinchinensis*
chumprak, *Heritiera cochinchinensis*
coachwood, *Ceratopetalum apetalum*
cocobolo, *Dalbergia retusa*
cocus, *Brya ebenus*
cocuswood, *Brya ebenus*
coigue, *Nothofagus dombeyi*
coihue, *Nothofagus dombeyi*
copalwood, Rhodesian, *Guibourtia coleosperma*
corail, *Pterocarpus soyauxii*
cordia, African, *Cordia* spp.
–, American light, *Cordia alliodora*
corkwood, *Ackama paniculata*
cornel, *Cornus florida*
cottonwood, *Populus balsamifera*
–, balm, *Populus trichocarpa*
–, balsam, *Populus trichocarpa*
–, black, *Populus trichocarpa*
–, eastern, *Populus deltoides*
–, Honduras, *Ceiba occidentalis*
–, western balm, *Populus trichocarpa*
–, western balsam, *Populus trichocarpa*
courbaril, *Astronium fraxinifolium*
–, *Hymenaea courbaril*
crabwood, *Carapa guianensis*

cudgerie, *Flindersia* spp.
–, maiden's blush, *Euroschinus falcatus*
cypress, *Liriodendron tulipifera*
–, *Podocarpus guatemalensis* §
–, bald, *Taxodium distichum* §
–, East African, *Cupressus* spp. §
–, Lawson's, *Chamaecyparis lawsoniana* §
–, Louisiana, *Taxodium distichum* §
–, nootka false, *Chamaecyparis nootkatensis* §
–, southern, *Taxodium distichum* §
–, swamp, *Taxodium distichum* §
–, yellow, *Chamaecyparis nootkatensis* §

dabéma, *Piptadeniastrum africanum*
dahoma, *Piptadeniastrum africanum*
dakua, *Agathis vitiensis* §
dalli, *Virola surinamensis*
damar minyak, *Agathis dammara* §
damoui, *Dracontomelum mangiferum*
damson, *Terminalia sericocarpa*
dang huong, *Pterocarpus pedatus*
daniellia, *Daniellia ogea*
danta, *Nesogordonia papaverifera*
dao, *Dracontomelum dao*
dau, *Dipterocarpus* spp.
deal, red, *Pinus sylvestris* §
–, white, *Picea abies* §
–, yellow, *Pinus sylvestris* §
–, yellow, Archangel, etc. ||, *Pinus sylvestris* §
degame, *Calycophyllum candidissimum*
den, *Abies* spp. §
denya, *Cylicodiscus gabunensis*
deodar, *Cedrus deodara* §
determa, *Ocotea rubra*
dhup, *Canarium euphyllum*
dhup, red, *Parishia insignis*
–, white, *Canarium euphyllum*
dibétou, *Lovoa trichilioides*
distemonanthus, *Distemonanthus benthamianus*
djave, *Baillonella toxisperma*
dogwood, *Cornus florida*
douka, *Tieghemella africana*
doussié, *Afzelia* spp.
duku, *Alstonia congensis*

eba, *Lophira alata*
ebben, *Diospyros* spp.
ébène, *Diospyros* spp.
Ebenholz, *Diospyros* spp.
ébiara, *Berlinia* spp.
ebony, African, *Diospyros* spp.
–, brown, *Brya ebenus*
–, Cameroons, *Diospyros* spp.
–, Ceylon, *Diospyros ebenum*
ebony, East Indian, *Diospyros ebenum*
–, Gaboon, *Diospyros* spp.
–, green, *Brya ebenus*
–, Jamaica, *Byra ebenus*
–, Kribi, *Diospyros* spp.
–, Macassar, *Diospyros celebica*
–, Madagascar, *Diospyros* spp.
–, Maracaibo, *Caesalpinia granadillo*
–, Mozambique, *Dalbergia melanoxylon*
–, Nigerian, *Diospyros* spp.
edinam, *Entandrophragma angolense*
Eiche, *Quercus* spp.
eik, *Quercus* spp.
ekhimi, *Piptadeniastrum africanum*
ekki, *Lophira alata*
ekpogoi, *Berlinia* spp.
elm, American, *Ulmus americana*
–, Canadian rock, *Ulmus thomasii*
–, cork, *Ulmus thomasii*
–, cork bark, *Ulmus thomasii*
–, – –, *Ulmus hollandica* var. *hollandica*
–, crow's foot, *Heritiera trifoliolata*
–, Dutch, *Ulmus hollandica* var. *hollandica*
–, English, *Ulmus procera*
–, Flemish, *Ulmus carpinifolia*
–, French, *Ulmus carpinifolia*
–, hickory, *Ulmus thomasii*
–, Japanese, *Ulmus* spp.
–, mountain, *Ulmus glabra*
–, nave, *Ulmus procera*
–, red, *Ulmus procera*
–, rock, *Ulmus thomasii*
–, Scotch, *Ulmus glabra*
–, smooth-leaved, *Ulmus carpinifolia*
–, soft, *Ulmus americana*
–, white, *Ulmus americana*
–, white, *Ulmus glabra*
–, wych, *Ulmus glabra*
els, *Alnus* spp.
embero, *Lovoa trichilioides*
embuia, *Phoebe porosa*
emeri, *Terminalia ivorensis*
emien, *Alstonia congensis*
eng, *Dipterocarpus tuberculatus*
épicea, *Picea* spp. §
erable, *Acer* spp.
erima, *Octomeles sumatrana*
erimado, *Ricinodendron heudelotii*
Erle, *Alnus* spp.
erun, *Erythrophleum suaveolens*
–, *Erythrophleum ivorense*
es, *Fraxinus* spp.
esa, *Celtis* spp.
Esche, *Fraxinus* spp.

keruing, Sarawak, *Dipterocarpus* spp.
keruwing, *Dipterocarpus* spp.
kévazingo, *Guibourtia demeusei*
khaya, *Khaya ivorensis*
kiaat, *Pterocarpus angolensis*
Kiefer, *Pinus* spp. §
kingwood, *Dalbergia cearensis*
–, *Astronium fraxinifolium*
Kirsche, *Prunus avium*
kirundu, *Antiaris toxicaria*
kitola, *Oxystigma oxyphyllum*
kokko, *Albizia lebbek*
kokrodua, *Pericopsis elata*
konara, *Quercus glandulifera*
kosipo, *Entandrophragma candollei*
kotibé, *Nesogordonia papaverifera*
koto, *Pterygota bequaertii*
krabak, *Anisoptera* spp.
krala, *Khaya anthotheca*
krappa, *Carapa guianensis*
kruen, *Dipterocarpus* spp.
kumbuk, *Terminalia arjuna*
kuraru, *Andira inermis*
kusia, *Nauclea diderrichii*
kwao, *Adina cordifolia*
kwarie, *Vochysia spp.*
kwila, *Intsia bijuga*
kwow, *Adina cordifolia*
kyenkyen, *Antiaris toxicaria*

laburnum, *Laburnum anagyroides*
lacewood, *Platanus hybrida*
Lagoswood, *Khaya ivorensis*
lamao, *Pentacme contorta*
lanan, *Shorea* spp. (light weight, pale red)
lancewood, *Oxandra lanceolata*
–, degame, *Calycophyllum candidissimum*
landosan, *Aningeria* spp.
larch, Dunkeld, *Larix eurolepis* §
–, eastern, *Larix laricina* §
–, Eastern Canadian, *Larix laricina* §
–, European, *Larix decidua* §
–, hybrid, *Larix eurolepis* §
–, Japanese, *Larix kaempferi* §
–, Siberian, *Larix russica* §
–, tamarack, *Larix laricina* §
–, western, *Larix occidentalis* §
Lärche, *Larix* spp. §
lariks, *Larix* spp. §
lauan, dark red, *Shorea* spp. (medium weight, dark red)
–, light red, *Shorea almon* (light weight, pale red)
–, Mindanao white, *Pentacme mindanensis*
–, red, *Shorea negrosensis* (medium weight, dark red)
lauan, white, *Parashorea malaanonan*
–, white, *Pentacme contorta*
–, white, *Pentacme mindanensis*
–, white, *Shorea almon* (light weight, pale red)
laup, *Dracontomelum mangiferum*
laurel, Chilean, *Laurelia sempervirens*
–, Ecuador, *Cordia alliodora*
–, Indian, *Terminalia alata*
–, native, *Cryptocarya glaucescens*
– de costa, *Cordia alliodora*
lemonwood, *Calycophyllum candidissimum*
letterwood, *Piratinera guianensis*
libengi, *Guibourtia arnoldiana*
lignum vitae, *Guaiacum* spp.
– –, Maracaibo, *Bulnesia arborea*
limba, *Terminalia superba*
limbo, *Terminalia superba*
lime, American, *Tilia americana*
–, English, *Tilia vulgaris*
–, European, *Tilia vulgaris*
–, Japanese, *Tilia japonica*
Linde, *Tilia* spp.
lingue, *Persea lingue*
lingué, *Afzelia africana*
locust, *Hymenaea courbaril*
–, black, *Robinia pseudoacacia*
–, West Indian, *Hymenaea courbaril*
locustwood, *Astronium fraxinifolium*
lolagbola, *Oxystigma oxyphyllum*
loliondo, *Olea welwitschii*
longleaf, *Pinus palustris* §
longui rouge, *Chrysophyllum cainito*
lotofa, *Sterculia rhinopetala*
loup, *Dracontomelum mangiferum*
louro, *Cordia trichotoma*
–, red, *Ocotea rubra*
– inamuhy, *Ocotea barcellensis*
– inamui, *Ocotea barcellensis*
– preto, *Ocotea* and *Nectandra* spp.
– vermelho, *Ocotea rubra*
lun, *Shorea* spp. (medium weight, yellow or white)
– kuning, *Shorea* spp. (medium weight, yellow)
– puteh, *Shorea* spp. (medium weight, white)
lunumidella, *Melia composita*
lup, *Dracontomelum mangiferum*

macula, *Eucalyptus maculata*
mafu, *Fagaropsis angolensis*
magnolia, *Magnolia* spp.
–, *Galboulimima baccata*
mahogany, brush, *Geissois benthami*
–, African, *Khaya ivorensis*
–, American, *Swietenia candollei*

meranti, red, *Shorea* spp. (light weight, pale red)
–, white, *Shorea* spp. (medium weight, white)
–, yellow, *Shorea* spp. (medium weight, yellow)
– bunga, *Shorea albida* (light weight, pale red)
– gerutu, *Parashorea* spp.
merawan, *Hopea* spp.
merbau, *Intsia bijuga*
merebong, *Dactylocladus stenostachys*
merisier, *Prunus avium*
mersawa, *Anisoptera* spp.
merubong, *Dactylocladus stenostachys*
messmate, *Eucalyptus cloeziana*
–, red, *Eucalyptus resinifera*, etc.
–, yellow, *Eucalyptus consideniana*
mfu, *Fagaropsis angolensis*
mfunda, *Cynometra webberi*
mgongo, *Ricinodendron rautanenii*
milkwood, *Alstonia scholaris*
minzu, *Combretodendron macrocarpum*
miro, *Podocarpus ferrugineus*§
missanda, *Erythrophleum suaveolens*
–, *Erythrophleum ivorense*
mjombo, *Brachystegia boehmii*
mkangazi, *Khaya nyasica*
mkora, *Afzelia quanzensis*
mlimangombe, *Rapanea rhododendroides*
mninga, *Pterocarpus angolensis*
moabi, *Baillonella toxisperma*
mogno, *Swietenia macrophylla(?)*
mora, *Mora excelsa*
moralfino, *Chlorophora tinctoria*
moreira, *Chlorophora excelsa*
moroti, *Balfourodendron riedelianum*
mpewere, *Piptadeniastrum africanum*
mpingo, *Dalbergia melanoxylon*
mringaringa, *Cordia abyssinica*
mtambara, *Cephalosphaera usambarensis*
mtondo, *Brachystegia spiciformis*
muave, *Erythrophleum ivorense*
mugaita, *Rapanea rhododendroides*
mugoma, *Cordia millenii*
mugongo, *Ricinodendron rautanenii*
muhimbi, *Cynometra alexandri*
muhindi, *Cynometra alexandri*
muhugwe, *Brachylaena hutchinsii*
muhuhu, *Brachylaena hutchinsii*
muirapiranga, *Brosimum paraense*
mujua, *Alstonia congensis*
mukarakati, *Fagaropsis angolensis*
mukebu, *Cordia millenii*
mukumari, *Cordia abyssinica*
mukushi, *Baikiaea plurijuga*
mukusi, *Baikiaea plurijuga*
mukwa, *Pterocarpus angolensis*
mulga, *Acacia aneura*
mumara, *Erythrophleum suaveolens*
muna, *Aningeria* spp.
mungongo, *Ricinodendron rautanenii*
muninga, *Pterocarpus angolensis*
munyama, *Khaya anthotheca*
muruma, *Fagaropsis angolensis*
musaisi, *Cassipourea malosana*
musheragi, *Olea hochstetteri*
musine, *Croton megalocarpus*
musizi, *Maesopsis eminii*
musodo, *Ricinodendron heudelotii*
mussacossa, *Afzelia quanzensis*
mussibi, *Guibourtia coleosperma*
mutenye, *Guibourtia arnoldiana*
mutti, *Terminalia alata*
muxibe, *Guibourtia coleosperma*
muzaule, *Guibourtia coleosperma*
mvule, *Chlorophora excelsa*
mwafu, *Canarium schweinfurthii*
myrtle, *Nothofagus cunninghamii*
–, Tasmanian, *Nothofagus cunninghamii*
mzimbe, *Cassipourea verticillata*

naga, *Brachystegia* spp.
nargusta, *Terminalia amazonia*
narra, *Pterocarpus indicus*
nazingu, *Mitragyna stipulosa*
ndiri, *Cassipourea malosana*
nemesu, *Shorea pauciflora* (medium weight)
ngale, *Poga oleosa*
ngollon, *Khaya ivorensis*
niangon, *Tarrietia utilis*
niove, *Staudtia stipitata*
nire, *Ulmus* spp.
njatuh, *Palaquium* spp.
nongo, *Albizia grandibracteata*
–, red, *Albizia grandibracteata*
–, white, *Albizia grandibracteata*
noyer, *Juglans* spp.
– d'Afrique, *Lovoa trichilioides*
– de Gabon, *Lovoa trichilioides*
– du Mayombe, *Terminalia superba*
nvero, *Lovoa trichilioides*
nyankom, *Tarrietia utilis*
nyatoh, *Palaquium* spp.
nzingu, *Mitragyna stipulosa*

oak, American red, *Quercus* spp.
–, – white, *Quercus* spp.
–, brown tulip, *Heritiera trifoliolata*
–, cherrybark, *Quercus falcata* var. *pagodaefolia*
–, chestnut, *Quercus prinus*
–, durmast, *Quercus petraea*

oak, English, *Quercus robur*
–, European, *Quercus robur*
–, evergreen, *Quercus ilex*
–, French, *Quercus robur*
–, holm, *Quercus ilex*
–, Japanese, *Quercus* spp.
–, New Zealand, *Beilschmiedia tawa*
–, northern red, *Quercus rubra*
–, overcup, *Quercus lyrata*
–, pedunculate, *Quercus robur*
–, Persian, *Quercus castaneaefolia*
–, Polish, *Quercus robur*
–, red tulip, *Heritiera peralata*
–, sessile, *Quercus petraea*
–, shumard red, *Quercus shumardii*
–, Slavonian, *Quercus robur*
–, southern red, *Quercus falcata* var. *falcata*
–, Spanish, *Quercus falcata* var. *falcata*
–, swamp chestnut, *Quercus michauxii*
–, swamp red, *Quercus falcata* var. *pagodaefolia*
–, Tasmanian, *Eucalyptus delegatensis*
–, Turkey, *Quercus cerris*
–, white, *Quercus alba*
oba suluk, *Shorea* spp. (medium weight, dark red)
obeche, *Triplochiton scleroxylon*
obobo, *Guarea cedrata*
obobonekwi, *Guarea thompsonii*
obobonufua, *Guarea cedrata*
ocote pine, *Pinus oocarpa* §
odoko, *Scottellia coriacea*
odum, *Chlorophora excelsa*
ofram, *Terminalia superba*
ofun, *Mansonia altissima*
ogea, *Daniellia ogea*
ogiovu, *Antiaris toxicaria*
ogoué, *Tarrietia densiflora*
ogwango, *Khaya ivorensis*
ohabu, *Pometia pinata*
ohia, *Celtis* spp.
ohnara, *Quercus mongolica*
okan, *Cylicodiscus gabunensis*
okoko, *Sterculia oblonga*
okoumé, *Aucoumea klaineana*
okuro, *Albizia* spp.
okwen, *Brachystegia* spp.
olive, East African, *Olea hochstetteri*
–, Elgon, *Olea welwitschii*
olivier, white, *Terminalia amazonia*
olivillo, *Aextoxicon punctatum*
olon, *Fagara heitzii*
– dur, *Fagara macrophylla*
– tendre, *Fagara heitzii*
olonvogo, *Fagara macrophylla*
omo, *Cordia platythyrsa*
omu, *Entandrophragma candollei*
onionwood, *Cassipourea verticillata*
opepe, *Nauclea diderrichii*
oriental wood, *Endiandra palmerstonii*
orme, *Ulmus* spp.
oro, *Antiaris toxicaria*
osan, *Aningeria* spp.
otie, *Pycnanthus angolensis*
otu, *Cleistopholis patens*
otutu, *Nesogordonia papaverifera*
ovangkol, *Guibourtia ehie*
ovoga, *Poga oleosa*
owawe, *Combretodendron macrocarpum*
oziya, *Daniellia ogea*

padang, *Palaquium* spp.
padauk, *Pterocarpus dalbergioides*
–, *Pterocarpus soyauxii*
–, African, *Pterocarpus soyauxii*
–, Andaman, *Pterocarpus dalbergioides*
–, Burma, *Pterocarpus macrocarpus*
paldao, *Dracontomelum dao*
palissander, *Dalbergia* spp.
palissandre, *Dalbergia* spp.
palmhout, *Buxus* spp.
palo de sangre, *Virola koschnyi*
palosapis, *Anisoptera* spp.
panga panga, *Millettia stuhlmannii*
papao, *Afzelia africana*
Pappel, *Populus* spp.
parapara, *Jacaranda copaia*
Para wood, *Caesalpinia echinata*
partridgewood, *Andira inermis*
–, *Caesalpinia granadillo*
patternwood, *Alstonia congensis*
pau marfim, *Balfourodendron riedelianum*
pear, *Pyrus communis*
pearwood, Nigerian, *Guarea cedrata*
pecan, *Carya* spp.
–, bitter, *Carya aquatica*
–, sweet, *Carya illinoensis*
penda, *Xanthostemon* spp.
pengiran, *Anisoptera* spp.
pepperwood, *Cinnamomum lautratii*
perawan, *Shorea* spp. (light weight, pale red)
Pernambuco wood, *Caesalpinia echinata*
peroba, red, *Aspidosperma* spp.
–, white, *Paratecoma peroba*
– amarella, *Paratecoma peroba*
– branca, *Paratecoma peroba*
– de campos, *Paratecoma peroba*
– rosa, *Aspidosperma* spp.
peterebi, *Cordia trichotoma*
petir, *Pseudosindora* spp.

podo, *Podocarpus* spp. §
poga, *Poga oleosa*
Poon, *Calophyllum tomentosum*
poplar, *Liriodendron tulipifera*
–, balm, *Populus balsamifera*
–, balsam, *Populus balsamifera*
–, black, *Populus balsamifera*
–, black, *Populus* spp.
–, black Italian, *Populus canadensis* var. *serotina*
–, Canadian, *Populus balsamifera*
–, Canadian, *Populus grandidentata*
–, European black, *Populus nigra*
–, grey, *Populus canescens*
–, pink, *Euroschinus falcatus*
–, tacamahac, *Populus balsamifera*
–, tulip, *Liriodendron tulipifera*
–, western balsam, *Populus trichocarpa*
–, white, *Populus alba*
–, yellow, *Liriodendron tulipifera*
populier, *Populus* spp.
possentrie, *Hura crepitans*
potrodom, *Erythrophleum ivorense*
–, *Erythrophleum suaveolens*
pradoo, *Pterocarpus macrocarpus*
prima vera, *Tabebuia donnellsmithii*
pterygota, African, *Pterygota bequaertii*
punah, *Tetramerista glabra*
purpleheart, *Peltogyne* spp.
pyinkado, *Xylia dolabriformis*
pyinma, *Lagerstroemia speciosa*
–, Andaman, *Lagerstroemia hypoleuca*

quandong, *Elaeocarpus* spp.
quaruba, *Vochysia* spp.

ramin, *Gonstylus macrophyllum*
– telur, *Gonystylus macrophyllum*
rapanea, *Rapanea rhododendroides*
rauli, *Nothofagus procera*
–, red, *Pinus sylvestris* §
redwood, *Pinus sylvestris* §
–, *Sequoia sempervirens* §
–, Andaman, *Pterocarpus dalbergioides*
–, Archangel etc. ‖, *Pinus sylvestris* §
–, Californian, *Sequoia sempervirens* §
–, Zambesi, *Baikiaea plurijuga*
rimu, *Dacrydium cupressinum* §
robinia, *Robinia pseudoacacia*
robusta, *Populus robusta*
Rosenholz, *Dalbergia* spp.
rosewood, *Dysoxylum fraseranum*
–, African, *Guibourtia demeusei*
–, Bahia, *Dalbergia nigra*
–, Brazilian, *Dalbergia nigra*
rosewood, East Indian, *Dalbergia latifolia*
–, Honduras, *Dalbergia stevensonii*
–, Indian, *Dalbergia latifolia*
–, New Guinea, *Pterocarpus indicus*
–, Rio, *Dalbergia nigra*
rose zebrano, *Berlinia* spp.

sain, *Terminalia alata*
sal, *Shorea robusta* (heavy, yellow-brown)
salmwood, *Cordia alliodora*
samba, *Triplochiton scleroxylon*
sandalwood, *Santalum album*
sandbox, *Hura crepitans*
sanga sanga, *Ricinodendron heudelotii*
sangre palo, *Virola koschnyi*
Santa Maria, *Calophyllum brasiliense* var. *rekoi*
sao, *Hopea odorata*
sapele, *Entandrophragma cylindricum*
–, heavy, *Entandrophragma candollei*
Sapelewood, *Entandrophragma cylindricum*
sapelli, *Entandrophragma cylindricum*
–, acajou, *Entandrophragma cylindricum*
sapin, *Abies* spp. §
sapodilla, *Manilkara zapota*
sassafras, *Doryphora sassafras*
sasswood, *Erythrophleum ivorense, suaveolens*
satin walnut, *Liquidambar styraciflua*
satinash, grey, *Cleistocalyx gustavioides*
satinay, red, *Syncarpia hillii*
satiné, *Brosimum paraense*
– rubané, *Brosimum paraense*
satinwood, African, *Fagara macrophylla*
–, Ceylon, *Chloroxylon swietenia*
–, East Indian, *Chloroxylon swietenia*
–, Jamaica, *Fagara flava*
–, Nigerian, *Distemonanthus benthamianus*
–, Nigerian yellow, *Distemonanthus benthamianus*
–, San Domingan, *Fagara flava*
–, West Indian, *Fagara flava*
saule, *Salix* spp.
selangan, *Hopea* spp.
–, red, *Shorea albida* (heavy, red)
– batu, *Shorea* spp. (heavy, yellow-brown)
– batu, red, *Shorea* spp. (heavy, red)
– batu merah, *Shorea* spp. (heavy, red)
– merah, *Shorea albida* (heavy, red)
sempilor, *Dacrydium elatum* §
sen, *Acanthopanax ricinifolius*
sepetir, *Pseudosindora palustris*
–, *Sindora* spp.
–, swamp, *Pseudosindora palustris*
– paya, *Pseudosindora palustris*
sequoia, *Sequoia sempervirens* §
seraya, dark red, *Shorea* spp. (medium weight)

seraya, light red, *Shorea* spp. (light weight, pale red)
–, red, *Shorea* spp. (light weight, pale red)
–, white, *Parashorea malaanonan*
–, yellow, *Shorea* spp. (medium weight, yellow)
– kuning, *Shorea* spp. (medium weight, yellow)
– merah, *Shorea* spp. (light weight, pale red)
seringawan, *Shorea albida* (heavy, red)
serrete, *Byrsonima coriacea,* var. *spicata*
sheoak, *Casuarina* spp.
shina, *Tilia japonica*
shinanoki, *Tilia japonica*
shisham, *Dalbergia sissoo*
sifou, *Albizia* spp.
silk cotton, *Ceiba pentandra*
silkwood, *Flindersia acuminata*
–, bolly or tarzali, *Cryptocarya oblata*
–, maple, *Flindersia brayleyana*
silky-oak, African, *Grevillea robusta*
–, Australian, *Cardwellia sublimis*
–, northern, *Cardwellia sublimis*
silver-grey wood, Indian, *Terminalia bialata*
simaruba, *Simaruba amara*
simarupa, *Simaruba amara*
sindru, *Alstonia congensis*
sipo, *Entandrophragma utile*
siris, *Albizia lebbek*
–, red, *Albizia toona*
sissoo, *Dalbergia sissoo*
snakewood, *Piratinera guianensis*
sobu, *Cleistopholis patens*
sompong, *Tetrameles nudiflora*
sovereign wood, *Terminalia sericocarpa*
spar, *Picea* spp. §
spruce, common, *Picea abies* §
–, black, *Picea mariana* §
–, eastern Canadian, *Picea* spp. §
–, Engelmann, *Picea engelmannii* §
–, European, *Picea abies* §
–, hemlock, *Tsuga canadensis* §
–, maritime, *Picea glauca* §
–, mountain, *Picea engelmannii* §
–, New Brunswick, *Picea glauca* §
–, Norway, *Picea abies* §
–, Nova Scotia, *Picea glauca* §
–, Quebec, *Picea glauca* §
–, red, *Picea rubens* §
–, Rocky Mountain, *Picea engelmannii* §
–, St. John, *Picea glauca* §
–, silver, *Picea sitchensis* §
–, Sitka, *Picea sitchensis* §
–, tideland, *Picea sitchensis* §
–, western white, *Picea glauca* §
–, white, *Picea glauca* §

stavewood, *Heritiera trifoliolata*
sterculia, brown, *Sterculia rhinopetala*
–, red, *Sterculia rhinopetala*
–, white, *Sterculia oblonga*
–, yellow, *Sterculia oblonga*
steud, *Terminalia amazonia*
stoolwood, *Alstonia congensis*
stringybark, messmate, *Eucalyptus obliqua*
–, red, *Eucalyptus resinifera,* etc.
–, red or mountain, *Eucalyptus macrorrhyncha*
–, white, *Eucalyptus engenioides,* etc.
–, white gum top, *Eucalyptus delegatensis*
–, yellow, *Eucalyptus acmenioides*
subaha, *Mitragyna ciliata*
sucupira, *Bowdichia* spp.
–, black, *Bowdichia* spp.
sugi, *Cryptomeria japonica* §
supa, *Sindora* spp.
sycamore, *Acer pseudoplatanus*
–, *Platanus occidentalis*
–, silver, *Cryptocarya glaucescens*
– plane, *Acer pseudoplatanus*

ta khian nu, *Anogeissus acuminata*
takien, *Hopea odorata*
tali, *Erythrophleum ivorense*
–, *Erythrophleum suaveolens*
tallowwood, *Eucalyptus microcorys*
tamalan, *Dalbergia oliveri*
tamarack, *Larix laricina* §
tamo, *Fraxinus mandshurica*
tangare, *Carapa guianensis*
tangile, *Shorea polysperma* (medium weight, dark red)
tarzali, *Cryptocarya oblata*
tasua, *Amoora polystachys*
tatajuba, *Chlorophora tinctoria*
taukkyan, *Terminalia alata*
taun, *Pometia pinata*
tawa, *Beilschmiedia tawa*
tchitola, *Oxystigma oxyphyllum*
teak, *Tectona grandis*
–, Australian, *Flindersia australis*
–, Borneo, *Intsia bijuga*
–, Guyana, *Dicorynia guianensis*
–, New Guinea, *Vitex cofassus*
–, Rhodesian, *Baikiaea plurijuga*
tea-tree, *Mellaleuca* spp.
tepa, *Laurelia philippiana*
terentang, *Campnosperma* spp.
tetraberlinia, *Tetraberlinia tubmaniana*
thingadu, *Parashorea stellata*
thingan, *Hopea odorata*
thitka, *Pentace burmanica*

General Index